Bayes' Rule With Python

A Tutorial Introduction to Bayesian Analysis

James V Stone

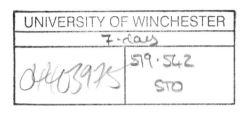
Title:
Bayes' Rule With Python
A Tutorial Introduction to Bayesian Analysis
Author: James V Stone
Published by Sebtel Press

First Edition, 2015.
Typeset in LaTeX 2_ε.
Cover Design by Stefan Brazzo.
Copyright ©2016 by James V Stone
Eighth printing.

ISBN 9780993367939

To mum.

It is remarkable that a science which began with the consideration of games of chance should have become the most important object of human knowledge.
Pierre Simon Laplace, 1812.

Contents

Preface

This introductory text is intended to provide a straightforward explanation of Bayes' rule, using plausible and accessible examples. It is written specifically for readers who have little mathematical experience, but who are nevertheless willing to acquire the required mathematics on a 'need to know' basis.

Lecturers (and authors) like to teach using a top-down approach, so they usually begin with abstract general principles, and then move on to more concrete examples. In contrast, students usually like to learn using a bottom-up approach, so they like to begin with examples, from which abstract general principles can then be derived. As this book is not designed to teach lecturers or authors, it has been written using a bottom-up approach. Accordingly, the first chapter contains several accessible examples of how Bayes' rule can be useful in everyday situations, and these examples are examined in more detail in later chapters. The reason for including many examples in this book is that, whereas one reader may grasp the essentials of Bayes' rule from a medical example, another reader may feel more comfortable with the idea of flipping a coin to find out if it is 'fair'. One side-effect of so many examples is that the book may appear somewhat repetitive. For this, I make no apology. As each example is largely self-contained, it can be read in isolation. The obvious advantages of this approach inevitably lead to a degree of repetition, but this is a small price to pay for the clarity that an introductory text should possess.

Computer Code in MatLab, Python and R

MatLab, Python and R code snippets can be downloaded from here:
jim-stone.staff.shef.ac.uk/BookBayes2012/BayesRuleMatlabCode.html
jim-stone.staff.shef.ac.uk/BookBayes2012/BayesRulePythonCode.html
jim-stone.staff.shef.ac.uk/BookBayes2012/BayesRuleRCode.html

Code Snippets Included in Text

This book contains exactly the same text as the book *Bayes' Rule: A Tutorial Introduction to Bayesian Analysis*, but also includes additional code snippets printed close to relevant equations and figures. For readers with some proficiency in programming, these snippets should aid understanding of the relevant equations.

The code snippets included within the text work on their own, but the corresponding online files contain additional lines which clean up the figures drawn by the code (e.g. by making plotted lines bold). These code snippets can be downloaded from the web site given above.

Corrections

Please email any corrections to j.v.stone@sheffield.ac.uk. A list of corrections is at http://jim-stone.staff.shef.ac.uk/BayesBook/Corrections.

Acknowledgments

Thanks to friends and colleagues for reading draft chapters, including David Buckley, Nikki Hunkin, Danielle Matthews, Steve Snow, Tom Stafford, Stuart Wilson, Paul Warren, Charles Fox, and to John de Pledge for suggesting the particular medical example in Section 2.6. Special thanks to Royston Sellman for providing most of the Python computer code, and to Patricia Revest for the R computer code. Thanks to those readers who have emailed me to point out errors. Finally, thanks to my wife, Nikki Hunkin, for sound advice on the writing of this book, during which she tolerated Bayesian reasoning being applied to almost every aspect of our lives.

Jim Stone,
Sheffield, England.

Chapter 1

An Introduction to Bayes' Rule

"... we balance probabilities and choose the most likely. It
is the scientific use of the imagination ... "
Sherlock Holmes, The Hound of the Baskervilles.
AC Doyle, 1901.

Introduction

Bayes' rule is a rigorous method for interpreting evidence in the context
of previous experience or knowledge. It was discovered by Thomas
Bayes (c. 1701-1761), and independently discovered by Pierre-Simon
Laplace (1749-1827). After more than two centuries of controversy,
during which Bayesian methods have been both praised and pilloried,
Bayes' rule has recently emerged as a powerful tool with a wide range

(a) Bayes (b) Laplace

Figure 1.1: The fathers of Bayes' rule. a) Thomas Bayes (c. 1701-1761).
b) Pierre-Simon Laplace (1749-1827).

of applications, which include: genetics[2], linguistics[12], image processing[15], brain imaging[33], cosmology[17], machine learning[5], epidemiology[26], psychology[31;44], forensic science[43], human object recognition[22], evolution[13], visual perception[23;41], ecology[32] and even the work of the fictional detective Sherlock Holmes[21]. Historically, Bayesian methods were applied by Alan Turing to the problem of decoding the German enigma code in the Second World War, but this remained secret until recently[16;29;37].

In order to appreciate the inner workings of any of the above applications, we need to understand why Bayes' rule is useful, and how it constitutes a mathematical foundation for reasoning. We will do this using a few accessible examples, but first, we will establish a few ground rules, and provide a reassuring guarantee.

Ground Rules

In the examples in this chapter, we will not delve into the precise meaning of probability, but will instead assume a fairly informal notion based on the frequency with which particular events occur. For example, if a bag contains 40 white balls and 60 black balls then the probability of reaching into the bag and choosing a black ball is the same as the proportion of black balls in the bag (ie 60/100=0.6). From this, it follows that the probability of an event (eg choosing a black ball) can adopt any value between zero and one, with zero meaning it definitely will not occur, and one meaning it definitely will occur. Finally, given a set of mutually exclusive events, such as the outcome of choosing a ball, which has to be either black or white, the probabilities of those events have to add up to one (eg 0.4+0.6=1). We explore the subtleties of the meaning of probability in Section 7.1.

A Guarantee

Before embarking on these examples, we should reassure ourselves with a fundamental fact regarding Bayes' rule, or *Bayes' theorem*, as it is also called: Bayes' theorem is not a matter of conjecture. By definition, a theorem is a mathematical statement that has been proved to be true. This is reassuring because, if we had to establish the rules for calculating with probabilities, we would insist that the result of such calculations must tally with our everyday experience of the physical world, just as surely as we would insist that $1 + 1 = 2$. Indeed, if we insist that probabilities must be combined with each other in accordance with certain common sense principles then Cox(1946)[7] showed that this leads to a unique set of rules, a set which includes Bayes' rule, which also appears as part of Kolmogorov's(1933)[24] (arguably, more rigorous) theory of probability.

1.1. Example 1: Poxy Diseases

The Patient's Perspective

Suppose that you wake up one day with spots all over your face, as in Figure 1.2. The doctor tells you that 90% of people who have smallpox have the same symptoms as you have. In other words, the probability of having these symptoms given that you have smallpox is 0.9 (ie 90%). As smallpox is often fatal, you are naturally terrified.

However, after a few moments of contemplation, you decide that you do not want to know the probability that you have these symptoms (after all, you already know you have them). Instead, what you really want to know is the probability that you have smallpox.

So you say to your doctor, "Yes, but what is the probability that I have smallpox given that I have these symptoms?". "Ah", says your doctor, "a very good question." After scribbling some equations, your doctor looks up. "The probability that you have smallpox given that you have these symptoms is 1.1%, or equivalently, 0.011." Of course,

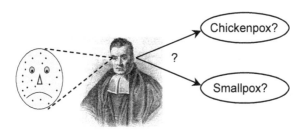

Figure 1.2: Thomas Bayes diagnosing a patient.

this is not good news, but it sounds better than 90%, and (more importantly) it is at least useful information. This demonstrates the stark contrast between the probability of the symptoms given a disease (which you do not want to know) and the probability of the disease given the symptoms (which you do want to know).

Bayes' rule transforms probabilities that look useful (but are often not) into probabilities that are useful. In the above example, the doctor used Bayes' rule to transform the uninformative probability of your symptoms given that you have smallpox into the informative probability that you have smallpox given your symptoms.

The Doctor's Perspective

Now, suppose you are a doctor, confronted with a patient who is covered in spots. The patient's symptoms are consistent with chickenpox, but they are also consistent with another, more dangerous, disease, smallpox. So you have a dilemma. You know that 80% of people with chickenpox have spots, but also that 90% of people with smallpox have spots. So the probability (0.8) of the symptoms given that the patient has chickenpox is similar to the probability (0.9) of the symptoms given that the patient has smallpox (see Figure 1.2).

If you were a doctor with limited experience then you might well think that both chickenpox and smallpox are equally probable. But, as you are a knowledgeable doctor, you know that chickenpox is common, whereas smallpox is rare. This knowledge, or *prior information*, can be used to decide which disease the patient probably has. If you had to guess (and you do have to guess because you are the doctor) then you would combine the possible diagnoses implied by the symptoms with your prior knowledge to arrive at a conclusion (ie that the patient probably has chickenpox). In order to make this example more tangible, let's run through it again, this time with numbers.

The Doctor's Perspective (With Numbers)

We can work out probabilities associated with a disease by making use of public health statistics. Suppose doctors are asked to report the number of cases of smallpox and chickenpox, and the symptoms observed. Using the results of such surveys, it is a simple matter to find the proportion of patients diagnosed with smallpox and chickenpox, and each patient's symptoms (eg spots). Using these data, we might find that the probability that a patient has spots given that the patient has smallpox is 90% or 0.9. We can write this in an increasingly succinct manner using a special notation

$$p(\text{symptoms are spots} \mid \text{disease is smallpox}) = 0.9, \qquad (1.1)$$

where the letter p stands for probability, and the vertical bar | stands for "given that". So, this short-hand statement should be read as

> "the probability that the patient's symptoms are spots given that he has smallpox is 90% or 0.9".

The vertical bar indicates that the probability that the patient has spots depends on the presence of smallpox. Thus, the probability of spots is said to be *conditional* on the disease under consideration. For this reason, such probabilities are known as *conditional probabilities*. We can write this even more succinctly as

$$p(\text{spots|smallpox}) = 0.9. \qquad (1.2)$$

Similarly, we might find that spots are observed in 80% of patients who have chickenpox, which is written as

$$p(\text{spots|chickenpox}) = 0.8. \qquad (1.3)$$

Equations 1.2 and 1.3 formalise why we should not use the symptoms alone to decide which disease the patient has. These equations take no account of our previous experience of the relative prevalence of smallpox and chickenpox, and are based only on the observed symptoms. As we shall see later, this is equivalent to making a decision based on the (in this case, false) assumption that both diseases are equally prevalent in the population, and that they are therefore *a priori* equally probable.

Note that the conditional probability $p(\text{spots|smallpox})$ is the probability of spots given that the patient has smallpox, but it is called the *likelihood* of smallpox (which is confusing, but standard, nomenclature). In this example, the disease smallpox has a larger likelihood than chickenpox. Indeed, as there are only two diseases under consideration, this means that, of the two possible alternatives, smallpox has the maximum likelihood. The disease with the maximum value of likelihood is known as the *maximum likelihood estimate* (MLE) of the disease that the patient has. Thus, in this case, the MLE of the disease is smallpox.

As discussed above, it would be hard to argue that we should disregard our knowledge or previous experience when deciding which disease the patient has. But exactly how should this previous experience be combined with current evidence (eg symptoms)? From a purely intuitive perspective, it would seem sensible to weight the likelihood of each disease according to previous experience of that disease, as in Figure 1.3. Since smallpox is rare, and is therefore intrinsically improbable, it might be sensible to weight the likelihood of smallpox by a small number. This would yield a small 'weighted likelihood', which would be a more realistic estimate of the probability that the patient has smallpox. For example, public health statistics may inform us that the prevalence of smallpox in the general population is 0.001, meaning that there is a one in a thousand chance that a

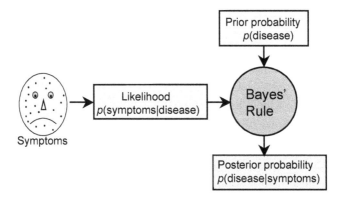

Figure 1.3: Schematic representation of Bayes' rule. Data, in the form of symptoms, are used find a likelihood, which is the probability of those symptoms given that the patient has a specific disease. Bayes' rule combines this likelihood with prior knowledge, and yields the posterior probability that the patient has the disease given that he has the symptoms observed.

randomly chosen individual has smallpox. Thus, the probability that a randomly chosen individual has smallpox is

$$p(\text{smallpox}) = 0.001. \tag{1.4}$$

This represents our prior knowledge about the disease in the population before we have observed our patient, and is known as the *prior probability* that any given individual has smallpox. As our patient (before we have observed his symptoms) is as likely as any other individual to have smallpox, we know that the prior probability that he has smallpox is 0.001.

If we follow our commonsense prescription, and simply weight (ie multiply) each likelihood by its prior probability then we obtain 'weighted likelihood' quantities which take account of the current evidence and of our prior knowledge of each disease. In short, this commonsense prescription leads to Bayes' rule. Even so, the equation for Bayes' rule given below is not obvious, and should be taken on trust for now. In the case of smallpox, Bayes' rule is

$$p(\text{smallpox}|\text{spots}) = \frac{p(\text{spots}|\text{smallpox}) \times p(\text{smallpox})}{p(\text{spots})}. \tag{1.5}$$

The term $p(\text{spots})$ in the denominator of Equation 1.5 is the proportion of people in the general population that have spots, and therefore represents the probability that a randomly chosen individual has spots. As will be explained on p16, this term is often disregarded, but we use a value that makes our sums come out neatly, and assume that $p(\text{spots}) = 0.081$ (ie 81 in every 1,000 individuals has spots). If we now substitute numbers into this equation then we obtain

$$p(\text{smallpox}|\text{spots}) \quad = \quad 0.9 \times 0.001/0.081 \tag{1.6}$$
$$= \quad 0.011, \tag{1.7}$$

which is the conditional probability that the patient has smallpox given that his symptoms are spots.

Crucially, the 'weighted likelihood' $p(\text{smallpox}|\text{spots})$ is also a conditional probability, but it is the probability of the disease smallpox given the symptoms observed, as shown in Figure 1.4. So, by making use of prior experience, we have transformed the conditional probability of the observed symptoms given a specific disease (the likelihood, which is based only on the available evidence) into a more useful conditional probability: the probability that the patient has a particular disease (smallpox) given that he has particular symptoms (spots).

In fact, we have just made use of Bayes' rule to convert one conditional probability, the likelihood $p(\text{spots}|\text{smallpox})$ into a more useful conditional probability, which we have been calling a 'weighted likelihood', but is formally known as the *posterior probability* $p(\text{smallpox}|\text{spots})$.

As noted above, both $p(\text{smallpox}|\text{spots})$ and $p(\text{spots}|\text{smallpox})$ are conditional probabilities, which have the same status from a mathematical viewpoint. However, for Bayes' rule, we treat them very differently.

Code Example 1.1: Smallpox
File: Ch1Eq06.py

```
# likelihood = prob of spots given smallpox
pSpotsGSmallpox = 0.9
# prior = prob of smallpox
pSmallpox = 0.001
# marginal likelihood = prob of spots
pSpots = 0.081
# find posterior = prob of smallpox given spots
pSmallpoxGSpots = pSpotsGSmallpox * pSmallpox / pSpots

print('Posterior, pSmallpoxGSpots = %.3f.' % pSmallpoxGSpots)
# Output: Posterior, pSmallpoxGSpots = 0.011.
```

The conditional probability $p(\text{spots}|\text{smallpox})$ is based only on the observed data (symptoms), and is therefore easier to obtain than the conditional probability we really want, namely $p(\text{smallpox}|\text{spots})$, which is also based on the observed data, but also on prior knowledge. For historical reasons, these two conditional probabilities have special names. As we have already seen, the conditional probability $p(\text{spots}|\text{smallpox})$ is the probability that a patient has spots given that he has smallpox, and is known as the likelihood of smallpox. The complementary conditional probability p(smallpox|spots) is the posterior probability that a patient has smallpox given that he has spots. In essence, Bayes' rule is used to combine prior experience (in the form of a prior probability) with observed data (spots) (in the form of a likelihood) to interpret these data (in the form of a posterior probability). This process is known as *Bayesian inference.*

Code Example 1.2: Chickenpox
File: Ch1Eq09.py

```
#   likelihood = prob of spots given chickenpox
pSpotsGChickenpox = 0.8
#   prior = prob of chickenpox
pChickenpox = 0.1
#   marginal likelihood = prob of spots
pSpots = 0.081
#   find posterior = prob of chickenpox given spots
pChickenpoxGSpots = pSpotsGChickenpox * pChickenpox / pSpots
print('Posterior, pChickenpoxGSpots = %.3f.' % pChickenpoxGSpots)

# Output:    Posterior, pChickenpoxGSpots = 0.988.
```

The Perfect Inference Engine

Bayesian inference is not guaranteed to provide the correct answer. Instead, it provides the probability that each of a number of alternative answers is true, and these can then be used to find the answer that is most probably true. In other words, it provides an informed guess. While this may not sound like much, it is far from random guessing. Indeed, it can be shown that no other procedure can provide a better guess, so that Bayesian inference can be justifiably interpreted as the output of a perfect guessing machine, a perfect inference engine (see Section 4.9, p100). This perfect inference engine is fallible, but it is provably less fallible than any other.

Making a Diagnosis

In order to make a diagnosis, we need to know the posterior probability of both of the diseases under consideration. Once we have both

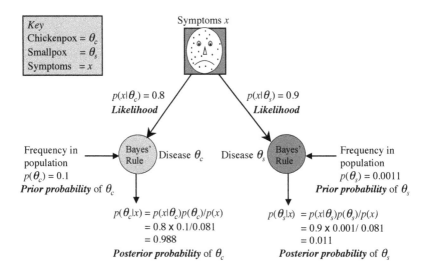

Figure 1.4: Comparing the probability of chickenpox and smallpox using Bayesian inference. The observed symptoms x seem to be more consistent with smallpox θ_s than chickenpox θ_c, as indicated by their likelihood values. However, the background rate of chickenpox in the population is higher than that of smallpox, which, in this case, makes it more probable that the patient has chickenpox, as indicated by its higher posterior probability.

posterior probabilities, we can compare them in order to choose the disease that is most probable given the observed symptoms.

Suppose that the prevalence of chickenpox in the general population is 10% or 0.1. This represents our prior knowledge about chickenpox before we have observed any symptoms, and is written as

$$p(\text{chickenpox}) = 0.1, \qquad (1.8)$$

which is the prior probability of chickenpox. As was done in Equation 1.6 for smallpox, we can weight the likelihood of chickenpox with its prior probability to obtain the posterior probability of chickenpox

$$
\begin{aligned}
p(\text{chickenpox}|\text{spots}) &= p(\text{spots}|\text{chickenpox}) \times p(\text{chickenpox})/p(\text{spots}) \\
&= 0.8 \times 0.1/0.081 \\
&= 0.988. \tag{1.9}
\end{aligned}
$$

The two posterior probabilities, summarised in Figure 1.4, are therefore

$$
\begin{aligned}
p(\text{smallpox}|\text{spots}) &= 0.011 \tag{1.10} \\
p(\text{chickenpox}|\text{spots}) &= 0.988. \tag{1.11}
\end{aligned}
$$

Thus, the posterior probability that the patient has smallpox is 0.011, and the posterior probability that the patient has chickenpox is 0.988. Aside from a rounding error, these sum to one.

Notice that we cannot be certain that the patient has chickenpox, but we can be certain that there is a 98.8% probability that he does. This is not only our best guess, but it is provably the best guess that can be obtained; it is effectively the output of a perfect inference engine.

In summary, if we ignore all previous knowledge regarding the prevalence of each disease then we have to use the likelihoods to decide which disease is present. The likelihoods shown in Equations 1.2 and 1.3 would lead us to diagnose the patient as probably having smallpox. However, a more informed decision can be obtained by taking account of prior information regarding the diseases under consideration. When we do take account of prior knowledge, Equations 1.10 and 1.11 indicate that the patient probably has chickenpox. In fact, these equations imply that the patient is about 89 (=0.988/0.011) times more likely to have chickenpox than smallpox. As we shall see later, this ratio of posterior probabilities plays a key role in Bayesian statistical analysis (Section 1.1, p14).

Taking account of previous experience yields the diagnosis that is most probable, given the evidence (spots). As this is the decision associated with the maximum value of the posterior probability, it is known as the *maximum a posteriori* or MAP estimate of the disease.

The equation used to perform Bayesian inference is called Bayes' rule, and in the context of diagnosis is

$$p(\text{disease}|\text{symptoms}) = \frac{p(\text{symptoms}|\text{disease})p(\text{disease})}{p(\text{symptoms})}, \qquad (1.12)$$

which is easier to remember as

$$\text{posterior} = \frac{\text{likelihood} \times \text{prior}}{\text{marginal likelihood}}. \qquad (1.13)$$

The *marginal likelihood* is also known as *evidence*, and we shall have more to say about it shortly.

Bayes' Rule: Hypothesis and Data

If we consider a putative disease to represent a specific hypothesis, and the symptoms to be some observed data then Bayes' rule becomes

$$p(\text{hypothesis}|\text{data}) = \frac{p(\text{data}|\text{hypothesis}) \times p(\text{hypothesis})}{p(\text{data})},$$

where the word "hypothesis" should be interpreted as, "hypothesis is true". Written in this form, the contrast between the likelihood and the posterior probability is more apparent. Specifically, the probability that the proposed hypothesis is true given some data that were actually observed is the posterior probability

$$p(\text{hypothesis}|\text{data}), \qquad (1.14)$$

whereas the probability of observing the data given that the hypothesis is true is the likelihood

$$p(\text{data}|\text{hypothesis}). \qquad (1.15)$$

A More Succinct Notation

We now introduce a succinct, and reasonably conventional, notation for the terms defined above. There is nothing new in the mathematics of this section, just a re-writing of equations used above. If we represent the observed symptoms by x, and the disease by the Greek letter *theta* θ_s (where the subscript s stands for smallpox) then we can write the conditional probability (ie the likelihood of smallpox) in Equation 1.2

$$p(x|\theta_s) = p(\text{spots}|\text{smallpox}) = 0.9. \tag{1.16}$$

Similarly, the background rate of smallpox θ_s in the population can be represented as the prior probability

$$p(\theta_s) = p(\text{smallpox}) = 0.001, \tag{1.17}$$

and the probability of the symptoms (the marginal likelihood) is

$$p(x) = p(\text{spots}) = 0.081. \tag{1.18}$$

Substituting this notation into Equation 1.5 (repeated here)

$$p(\text{smallpox}|\text{spots}) = \frac{p(\text{spots}|\text{smallpox}) \times p(\text{smallpox})}{p(\text{spots})}, \tag{1.19}$$

yields

$$p(\theta_s|x) = \frac{p(x|\theta_s) \times p(\theta_s)}{p(x)}. \tag{1.20}$$

Similarly, if we define

$$
\begin{aligned}
p(x|\theta_c) &= p(\text{spots}|\text{chickenpox}) \\
p(\theta_c|x) &= p(\text{chickenpox}|\text{spots}) \\
p(\theta_c) &= p(\text{chickenpox}),
\end{aligned}
\tag{1.21}
$$

then we can re-write Equation 1.9 to obtain the posterior probability of chickenpox as

$$p(\theta_c|x) = \frac{p(x|\theta_c) \times p(\theta_c)}{p(x)}. \tag{1.22}$$

If we use θ without a subscript to represent any disease (or hypothesis), and x to represent any observed symptoms (or data) then Bayes' rule can be written as (we now drop the use of the $\times$ symbol)

$$p(\theta|x) = \frac{p(x|\theta)p(\theta)}{p(x)}. \tag{1.23}$$

Finally, we should note that smallpox made history by being the first disease to be eradicated from the Earth in 1979, which makes the prior probability of catching it somewhat less than the value $p(\theta_s) = 0.001$ assumed in the above example.

Parameters and Variables: Notice that there is nothing special about which symbol stands for disease and which for symptoms, and that we could equally well have used θ to represent symptoms, and x to represent diseases. However, it is common to use a Greek letter like θ to represent the thing we wish to estimate, and x to represent the evidence (eg symptoms) on which our estimated value of θ will be based. Similarly, using an equally arbitrary but standard convention, the symbol that represents the thing we wish to estimate is usually called a *parameter* (θ), whereas the evidence used to estimate that thing is usually called a *variable* (x).

Model Selection, Posterior Ratios and Bayes Factors

As noted above, when we take account of prior knowledge, it turns out that the patient is about 90 times more likely (ie 0.988 vs 0.011) to have chickenpox than smallpox. Indeed, it is often the case that we wish to compare the relative probabilities of two hypotheses (eg diseases). As each hypothesis acts as a (simple) model for the data, and we wish to select the most probable model, this is known as *model selection*, which involves a comparison using a ratio of posterior probabilities.

The posterior ratio, which is also known as the *posterior odds* between the hypotheses θ_c and θ_s, is

$$R_{post} = \frac{p(\theta_c|x)}{p(\theta_s|x)}. \tag{1.24}$$

If we apply Bayes' rule to the numerator and denominator then

$$R_{post} \;=\; \frac{p(x|\theta_c)p(\theta_c)/p(x)}{p(x|\theta_s)p(\theta_s)/p(x)}, \tag{1.25}$$

where the marginal likelihood $p(x)$ cancels, so that

$$R_{post} \;=\; \frac{p(x|\theta_c)}{p(x|\theta_s)} \times \frac{p(\theta_c)}{p(\theta_s)}. \tag{1.26}$$

This is a product of two ratios, the ratio of likelihoods, or *Bayes factor*

$$B \;=\; \frac{p(x|\theta_c)}{p(x|\theta_s)}, \tag{1.27}$$

and the ratio of priors, or *prior odds* between θ_c and θ_s, which is

$$R_{prior} \;=\; p(\theta_c)/p(\theta_s). \tag{1.28}$$

Thus, the posterior odds can be written as

$$R_{post} \;=\; B \times R_{prior}, \tag{1.29}$$

which, in words, is: *posterior odds = Bayes factor × prior odds.* In this example, we have

$$R_{post} \;=\; \frac{0.80}{0.90} \times \frac{0.1}{0.001} = 88.9.$$

```
pSpotsGSmallpox = 0.9
pSmallpox = 0.001
pSpotsGChickenpox = 0.8
pChickenpox = 0.1
Rpost = pSpotsGChickenpox*pChickenpox / (pSpotsGSmallpox*pSmallpox)
print('Rpost = %.3f.' % Rpost)

# Output:   Rpost = 88.9
```

Note that the likelihood ratio (Bayes factor) is less than one (and so favours θ_s), whereas the prior odds is much greater than one (and favours θ_c), with the result that the posterior odds come out massively in favour of θ_c. If the posterior odds is greater than 3 or less than $1/3$ (in both cases one hypothesis is more than 3 times more probable than the other) then this is considered to represent a substantial difference between the probabilities of the two hypotheses[19], so a posterior odds of 88.9 is definitely substantial.

Ignoring the Marginal Likelihood

As promised, we consider the marginal likelihood p(symptoms) or $p(x)$ briefly here (and in Chapter 2 and Section 4.5). The marginal likelihood refers to the probability that a randomly chosen individual has the symptoms that were actually observed, which we can interpret as the prevalence of spots in the general population.

Crucially, the decision as to which disease the patient has depends only on the relative sizes of different posterior probabilities (eg Equations 1.10, 1.11, and in Equations 1.20,1.22). Note that each of these posterior probabilities is proportional to $1/p$(symptoms) in Equations 1.10, 1.11, also expressed as $1/p(x)$ in Equations 1.20,1.22. This means that a different value of the marginal probability p(symptoms) would change all of the posterior probabilities by the same proportion, and therefore has no effect on their *relative* magnitudes. For example, if we arbitrarily decided to double the value of the marginal likelihood from 0.081 to 0.162 then both posterior probabilities would be halved (from 0.011 and 0.988 to about 0.005 and 0.494), but the posterior probability of chickenpox would still be 88.9 times larger than the posterior probability of smallpox. Indeed, the previous section on Bayes factors relies on the fact that the ratio of two posterior probabilities is independent of the value of the marginal probability.

In summary, the value of the marginal probability has no effect on which disease yields the largest posterior probability (eg Equations 1.10 and 1.11), and therefore has no effect on the decision regarding which disease the patient probably has.

1.2. Example 2: Forkandles

The example above is based on medical diagnosis, but Bayes' rule can be applied to any situation where there is uncertainty regarding the value of a measured quantity, such as the acoustic signal that reaches the ear when some words are spoken. The following example follows a similar line of argument as the previous one, and aside from the change in context, provides no new information for the reader to absorb.

If you walked into a hardware store and asked, *Have you got fork handles?*, then you would be surprised to be presented with four candles. Even though the phrases *fork handles* and *four candles* are acoustically almost identical, the shop assistant knows that he sells many more candles than fork handles (Figure 1.5). This in turn, means that he probably does not even hear the words *fork handles*, but instead hears *four candles*. What has this got to do with Bayes' rule?

The acoustic data that correspond to the sounds spoken by the customer are equally consistent with two interpretations, but the assistant assigns a higher weighting to one interpretation. This weighting is based on his prior experience, so he knows that customers are more likely to request four candles than fork handles. The experience of the assistant allows him to hear what was probably said by the customer, even though the acoustic data was pretty ambiguous. Without knowing it, he has probably used something like Bayes' rule to hear what the customer probably said.

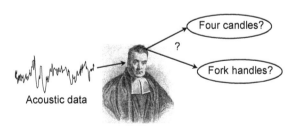

Figure 1.5: Thomas Bayes trying to make sense of a London accent, which removes the *h* sound from the word *handle*, so the phrase *fork handles* is pronounced *fork 'andles*, and therefore sounds like *four candles* (see Fork Handles YouTube clip by The Two Ronnies).

Likelihood: Answering the Wrong Question

Given that the two possible phrases are *four candles* and *fork handles*, we can formalise this scenario by considering the probability of the acoustic data given each of the two possible phrases. In both cases, the probability of the acoustic data depends on the words spoken, and this dependence is made explicit as two probabilities:

1) the probability of the acoustic data given *four candles* was spoken,
2) the probability of the acoustic data given *fork handles* was spoken.

A short-hand way of writing these is

$$p(\text{acoustic data}|\text{four candles})$$

$$p(\text{acoustic data}|\text{fork handles}), \qquad (1.30)$$

where the expression $p(\text{acoustic data}|\text{four candles})$, for example, is interpreted as the likelihood that the phrase spoken was *four candles*. As both phrases are consistent with the acoustic data, the probability of the data is almost the same in both cases. That is, the probability of the data given that *four candles* was spoken is almost the same as the probability of the data given that *fork handles* was spoken. For simplicity, we will assume that these probabilities are

$$p(\text{data}|\text{four candles}) = 0.6$$

$$p(\text{data}|\text{fork handles}) = 0.7. \qquad (1.31)$$

Knowing these two likelihoods does allow us to find an answer, but it is an answer to the wrong question. Each likelihood above provides an answer to the (wrong) question: *what is the probability of the observed acoustic data given that each of two possible phrases was spoken?*

Posterior Probability: Answering the Right Question

The right question, the question to which we would like an answer is: what is the probability that each of the two possible phrases was spoken given the acoustic data? The answer to this, the right question, is implicit in two new conditional probabilities, the posterior probabilities

$$p(\text{four candles}|\text{data})$$
$$p(\text{fork handles}|\text{data}), \qquad (1.32)$$

as shown in Figures 1.6 and 1.7. Notice the subtle difference between the pair of Equations 1.31 and the pair 1.32. Equations 1.31 tells us the likelihoods, the probability of the data given two possible phrases, which turn out to be almost identical for both phrases in this example. In contrast, Equations 1.32 tells us the posterior probabilities, the probability of each phrase given the acoustic data.

Crucially, each likelihood tells us the probability of the data given a particular phrase, but takes no account of how often that phrase has been given (ie has been encountered) in the past. In contrast, each posterior probability depends, not only on the data (in the form of the likelihood), but also on how frequently each phrase has been encountered in the past; that is, on prior experience.

So, we want the posterior probability, but we have the likelihood. Fortunately, Bayes' rule provides a means of getting from the likelihood to the posterior, by making use of extra knowledge in the form of prior experience, as shown in Figure 1.6.

Prior Probability

Let's suppose that the assistant has been asked for four candles a total of 90 times in the past, whereas he has been asked for fork handles only 10 times. To keep matters simple, let's also assume that the next customer will ask either for four candles or fork handles (we will revisit this simplification later). Thus, before the customer has uttered a single word, the assistant estimates that the probability that he will say each of the two phrases is

$$p(\text{four candles}) \quad = \quad 90/100 = 0.9$$
$$p(\text{fork handles}) \quad = \quad 10/100 = 0.1. \qquad (1.33)$$

These two prior probabilities represent the prior knowledge of the assistant, based on his previous experience of what customers say.

When confronted with an acoustic signal that has one of two possible interpretations, the assistant naturally interprets this as *four candles*, because, according to his past experience, this is what such ambiguous acoustic data usually means in practice. So, he takes the two almost equal likelihood values, and assigns a weighting to each one, a weighting that depends on past experience, as in Figure 1.7. In other words, he uses the acoustic data, and combines it with his previous experience to make an inference about which phrase was spoken.

Inference

One way to implement this weighting (ie to do this inference) is to simply multiply the likelihood of each phrase by how often that phrase has occurred in the past. In other words, we multiply the likelihood of each putative phrase by its corresponding prior probability. The result

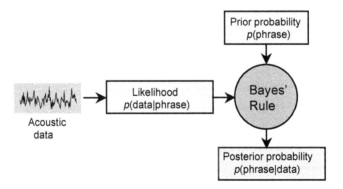

Figure 1.6: A schematic representation of Bayes' rule. Data alone, in the form of acoustic data, can be used to find a likelihood value, which is the conditional probability of the acoustic data given some putative spoken phrase. When Bayes' rule is used to combine this likelihood with prior knowledge then the result is a posterior probability, which is the probability of the phrase given the observed acoustic data.

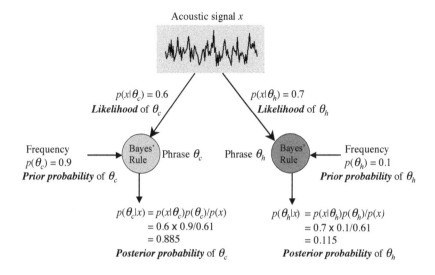

Figure 1.7: Bayesian inference applied to speech data.

yields a posterior probability for each possible phrase

$$p(\text{four candles}|\text{data}) = \frac{p(\text{data}|\text{four candles})p(\text{four candles})}{p(\text{data})}$$

$$p(\text{fork handles}|\text{data}) = \frac{p(\text{data}|\text{fork handles})p(\text{fork handles})}{p(\text{data})}, \quad (1.34)$$

where $p(\text{data})$ is the marginal likelihood, which is the probability of the observed data.

In order to ensure that the posterior probabilities sum to one, the value of $p(\text{data})$ is 0.61 in this example, but as we already know from Section 1.1 (p16), its value is not important for our purposes. If we substitute the likelihood and prior probability values defined in Equations 1.31 and 1.33 in 1.34 then we obtain their posterior probabilities as

$$
\begin{aligned}
p(\text{four candles}|\text{data}) &= p(\text{data}|\text{four candles})p(\text{four candles})/p(\text{data}) \\
&= 0.6 \times 0.9/0.61 = 0.885,
\end{aligned}
$$

$$
\begin{aligned}
p(\text{fork handles}|\text{data}) &= p(\text{data}|\text{fork handles})p(\text{fork handles})/p(\text{data}) \\
&= 0.7 \times 0.1/0.61 = 0.115.
\end{aligned}
$$

As in the previous example, we can write this more succinctly by defining

$$
\begin{aligned}
x &= \text{acoustic data,} \\
\theta_c &= \text{four candles,} \\
\theta_h &= \text{fork handles,}
\end{aligned}
$$

so that

$$
\begin{aligned}
p(\theta_c|x) &= p(x|\theta_c)p(\theta_c)/p(x) = 0.885 \\
p(\theta_h|x) &= p(x|\theta_h)p(\theta_h)/p(x) = 0.115. \quad\quad (1.35)
\end{aligned}
$$

Code Example 1.4: Four Candles
File: Ch1Eq34.py

```
pData = 0.61
pDataGFourCandles = 0.6
pFourCandles = 0.9
pFourCandlesGData = pDataGFourCandles * pFourCandles / pData
print 'pFourCandlesGData␣=␣%.3f' % pFourCandlesGData

pDataGForkHandles = 0.7
pForkHandles = 0.1
pForkHandlesGData = pDataGForkHandles * pForkHandles / pData
print 'pForkHandlesGData␣=␣%.3f\n' % pForkHandlesGData

# Output:    pFourCandlesGData = 0.885
#            pForkHandlesGData = 0.115
```

These two posterior probabilities represent the answer to the right question, so we can now see that the probability that the customer said *four candles* is 0.885 whereas the probability that the customer said *fork handles* was 0.115. As *four candles* is associated with the highest value of the posterior probability, it is the *maximum a posteriori* (MAP) estimate of the phrase that was spoken. The process that makes use of evidence (symptoms) to produce these posterior probabilities is called *Bayesian inference*.

1.3. Example 3: Flipping Coins

This example follows the same line of reasoning as those above, but also contains specific information on how to combine probabilities from independent events, such as coin flips. This will prove crucial in a variety of contexts, and in examples considered later in this book.

Here, our task is to decide how unfair a coin is, based on just two coin flips. Normally, we assume that coins are fair or unbiased, so that a large number of coin flips (eg 1000) yields an equal number of heads and tails. But suppose there was a fault in the machine that minted coins, so that each coin had more metal on one side or the other, with the result that each coin is biased to produce more heads than tails, or vice versa. Specifically, 25% of the coins produced by the machine have a bias of 0.4, and 75% have a bias of 0.6. By definition, a coin with a bias of 0.4 produces a head on 40% of flips, whereas a coin with a bias of 0.6 produces a head on 60% of flips (on average). Now, suppose we choose one coin at random, and attempt to decide which of the two bias

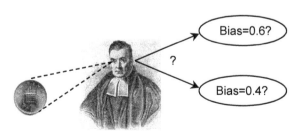

Figure 1.8: Thomas Bayes trying to decide the value of a coin's bias.

values it has. For brevity, we define the coin's bias with the parameter θ, so the true value of θ for each coin is either $\theta_{0.4} = 0.4$, or $\theta_{0.6} = 0.6$.

One Coin Flip: Here we use one coin flip to define a few terms that will prove useful below. For each coin flip, there are two possible outcomes, a head x_h, and a tail x_t. For example, if the coin's bias is $\theta_{0.6}$ then, by definition, the conditional probability of observing a head is $\theta_{0.6}$

$$p(x_h|\theta_{0.6}) = \theta_{0.6} = 0.6. \qquad (1.36)$$

Similarly, the conditional probability of observing a tail is

$$p(x_t|\theta_{0.6}) \quad = \quad (1 - \theta_{0.6}) = 0.4, \qquad (1.37)$$

where both of these conditional probabilities are likelihoods. Note that we follow the convention of the previous examples by using θ to represent the parameter whose value we wish to estimate, and x to represent the data used to estimate the true value of θ.

Two Coin Flips: Consider a coin with a bias θ (where θ could be 0.4 or 0.6, for example). Suppose we flip this coin twice, and obtain a head x_h followed by a tail x_t, which define the ordered list or *permutation*

$$\mathbf{x} \quad = \quad (x_h, x_t). \qquad (1.38)$$

As the outcome of one flip is not affected by any other flip outcome, outcomes are said to be *independent* (see Section 2.2 or Appendix C). This independence means that the probability of observing any two

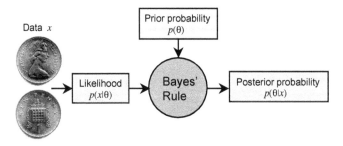

Figure 1.9: A schematic representation of Bayes' rule, applied to the problem of estimating the bias of a coin based on two coin flips.

outcomes can be obtained by multiplying their probabilities

$$p(\mathbf{x}|\theta) \quad = \quad p((x_h, x_t)|\theta) \tag{1.39}$$

$$= \quad p(x_h|\theta) \times p(x_t|\theta). \tag{1.40}$$

More generally, for a coin with a bias θ, the probability of a head x_h is $p(x_h|\theta) = \theta$, and the probability of a tail x_t is therefore $p(x_t|\theta) = (1-\theta)$. It follows that Equation 1.40 can be written as

$$p(\mathbf{x}|\theta) \quad = \quad \theta \times (1 - \theta), \tag{1.41}$$

which will prove useful below.

The Likelihoods of Different Coin Biases: According to Equation 1.41, if the coin bias is $\theta_{0.6}$ then

$$p(\mathbf{x}|\theta_{0.6}) \quad = \quad \theta_{0.6} \times (1 - \theta_{0.6}) \tag{1.42}$$
$$= \quad 0.6 \times 0.4 \tag{1.43}$$
$$= \quad 0.24, \tag{1.44}$$

and if the coin bias is $\theta_{0.4}$ then (the result is the same)

$$p(\mathbf{x}|\theta_{0.4}) \quad = \quad \theta_{0.4} \times (1 - \theta_{0.4}) \tag{1.45}$$
$$= \quad 0.4 \times 0.6 \tag{1.46}$$
$$= \quad 0.24. \tag{1.47}$$

Note that the only difference between these two cases is the reversed ordering of terms in Equations 1.43 and 1.46, so that both values of θ have equal likelihood values. In other words, the observed data $\mathbf{x}$ are equally probable given the assumption that $\theta_{0.4} = 0.4$ or $\theta_{0.6} = 0.6$, so they do not help in deciding which bias our chosen coin has.

Prior Probabilities of Different Coin Biases: We know (from above) that 25% of all coins have a bias of $\theta_{0.4}$, and that 75% of all coins have a bias of $\theta_{0.6}$. Thus, even before we have chosen our coin, we know (for example) there is a 75% chance that it has a bias of 0.6. This information defines the prior probability that any coin has one of two bias values, either $p(\theta_{0.4}) = 0.25$, or $p(\theta_{0.6}) = 0.75$.

Posterior Probabilities of Different Coin Biases: As in previous examples, we adopt the naïve strategy of simply weighting each likelihood value by its corresponding prior (and dividing by $p(\mathbf{x})$) to obtain Bayes' rule

$$
\begin{aligned}
p(\theta_{0.4}|\mathbf{x}) &= p(\mathbf{x}|\theta_{0.4})p(\theta_{0.4})/p(\mathbf{x}) \\
&= 0.24 \times 0.25/0.24 \\
&= 0.25, \quad\quad\quad\quad\quad\quad (1.48) \\
p(\theta_{0.6}|\mathbf{x}) &= p(\mathbf{x}|\theta_{0.6})p(\theta_{0.6})/p(\mathbf{x}) \\
&= 0.24 \times 0.75/0.24 \\
&= 0.75. \quad\quad\quad\quad\quad\quad (1.49)
\end{aligned}
$$

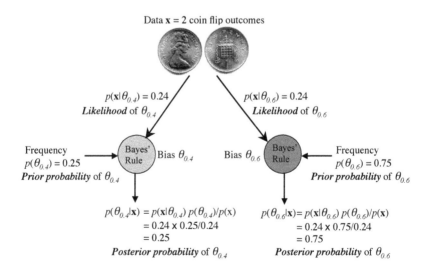

Figure 1.10: Bayesian inference applied to coin flip data.

Figure 1.11: Is this a hill or a crater? Try turning the book upside-down. (Barringer crater, with permission, United States Geological Survey).

In order to ensure posterior probabilities sum to one, we have assumed a value for the marginal probability of $p(\mathbf{x}) = 0.24$ (but we know from p16 that its value makes no difference to our final decision about coin bias). As shown in Figures 1.9 and 1.10, the probabilities in Equations 1.48 and 1.49 take account of both the data and of prior experience, and are therefore posterior probabilities. In summary, whereas the equal likelihoods in this example (Equations 1.44 and 1.47) did not allow us to choose between the coin biases $\theta_{0.4}$ and $\theta_{0.6}$, the values of the posterior probabilities (Equations 1.48 and 1.49) imply that a bias of $\theta_{0.6}$ is 3 (=0.75/0.25) times more probable than a bias is $\theta_{0.4}$.

1.4. Example 4: Light Craters

When you look at Figure 1.11, do you see a hill or a crater? Now turn the page upside-down. When you invert the page, the content of the picture does not change, but what you see does change (from a hill to a crater). This illusion almost certainly depends on the fact that your visual system assumes that the scene is lit from above. This, in turn, forces you to interpret the Figure 1.11 as a hill, and the inverted version as a crater (which it is, in reality).

In terms of Bayes' rule, the image data are equally consistent with a hill and a crater, where each interpretation corresponds to a different maximum likelihood value. Therefore, in the absence of any prior assumptions on your part, you should see the image as depicting either a hill or a crater with equal probability. However, the assumption that light comes from above corresponds to a prior, and this effectively forces you to interpret the image as a hill or a crater, depending on whether the image is inverted or not. Note that there is no uncertainty or noise; the image is perfectly clear, but also perfectly ambiguous without the addition of a prior regarding the light source. This example demonstrates that Bayesian inference is useful even when there is no noise in the observed data, and that even the apparently simple act of seeing requires the use of prior information [10;40;41;42]:

> Seeing is not a direct apprehension of reality, as we often like to pretend. Quite the contrary: *seeing is inference from incomplete information ...*
> ET Jaynes, 2003 (p133) [18].

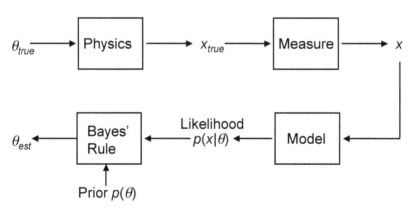

Figure 1.12: Forward and inverse probability.
Top: Forward probability. A parameter value θ_{true} (eg coin bias) which is implicit in a physical process (eg coin flipping) yields a quantity x_{true} (eg proportion of heads), which is measured as x, using an imperfect measurement process.
Bottom: Inverse probability. Given a mathematical model of the physics that generated x_{true}, the measured value x implies a range of possible values for the parameter θ. The probability of x given each possible value of θ defines a likelihood. When combined with a prior, each likelihood yields a posterior probability $p(\theta|x)$, which allows an estimate θ_{est} of θ_{true} to be obtained.

1.5. Forward and Inverse Probability

If we are given a coin with a known bias of say, $\theta = 0.6$, then the probability of a head for each coin flip is given by the likelihood $p(x_h|\theta) = 0.6$. This is an example of a *forward probability*, which involves calculating the probability of each of a number of different consequences (eg obtaining two heads) given some known cause or fact, see Figure 1.12. If this coin is flipped a 100 times then the number of heads could be 62, so the actual proportion of heads is $x_{true} = 0.62$. But, because no measurement is perfectly reliable, we may mis-count 62 as 64 heads, so the measured proportion is $x = 0.64$. Consequently, there is a difference, often called *noise*, between the true coin bias and the measured proportion of heads. The source of this noise may be due to the probabilistic nature of coin flips or to our inability to measure the number of heads accurately. Whatever the cause of the noise, the only information we have is the measured number of heads, and we must use this information as wisely as possible.

The converse of reasoning forwards from a given physical parameter or scenario involves a harder problem, also illustrated in Figure 1.12. Reasoning backwards from measurements (eg coin flips or images) amounts to finding the posterior or *inverse probability* of the value of an unobserved variable (eg coin bias, 3D shape), which is usually the cause of the observed measurement. By analogy, arriving at the scene of a crime, a detective must reason backwards from the clues, as eloquently expressed by Sherlock Holmes:

> Most people, if you describe a train of events to them, will tell you what the result would be. They can put those events together in their minds, and argue from them that something will come to pass. There are few people, however, who, if you told them a result, would be able to evolve from their own inner consciousness what the steps were that led to that result. This power is what I mean when I talk of reasoning backward, or analytically.
>
> Sherlock Holmes, from A Study in Scarlet. AC Doyle, 1887.

Indeed, finding inverse probabilities is precisely the problem Bayes' rule is designed to tackle.

Summary

All decisions should be based on evidence, but the best decisions should also be based on previous experience. The above examples demonstrate not only that prior experience is crucial for interpreting evidence, but also that Bayes' rule provides a rigorous method for doing so.

Chapter 2

Bayes' Rule in Pictures

> Probability theory is nothing but common sense reduced to calculation.
> Pierre-Simon Laplace (1749-1827).

Introduction

Some people understand mathematics through the use of symbols, but additional insights are often obtained if those symbols can be translated into geometric diagrams and pictures. In this chapter, once we have introduced random variables and the basic rules of probability, several different pictorial representations of probability will be used to encourage an intuitive understanding of the logic that underpins those rules. Having established a firm understanding of these rules, Bayes' rule follows using a few lines of algebra.

2.1. Random Variables

As in the previous chapter, we define the bias of a coin as its propensity to land heads up. But now we are familiar with the idea of quantities (such as coin flip outcomes) that are subject to variability, we can consider the concept of a *random variable*. The term random variable continues in use for historical reasons, although random variables are not the same as the variables used in algebra, like the x in $3x + 2 = 5$, where the variable x has a definite, but unknown, value that we can solve for.

Each value of a random variable can be considered as one possible outcome of an experiment that has a number of different possible outcomes. If the experiment is, say, 5 flips of a coin then we obtain a sequence of 5 heads and tails, for example, $(x_h, x_h, x_t, x_t, x_h)$. We could choose to treat this sequence as the outcome of the experiment; alternatively, we could treat the number of heads in the sequence as the outcome. In this case, the random variable X (written in upper case by convention) has six possible values between 0 and 5, using a standard notation for a *set* of items

$$X = \{0, 1, 2, 3, 4, 5\}, \tag{2.1}$$

where this set is known as the *sample space* of X. The probability that $X = 3$, for example, is written as $p(X = 3)$. More commonly, this is represented with the variables $x_0, x_1, \ldots$

$$X = \{x_0, x_1, x_2, x_3, x_4, x_5\}, \tag{2.2}$$

where the probability that $X = x_3$, for example, is written as $p(X = x_3)$ or just $p(x_3)$. When considered over all values of X, the probability associated with each value of X defines the set of probabilities that comprise the *probability distribution*

$$p(X) = \{ p(x_0),\ p(x_1),\ p(x_2),\ p(x_3),\ p(x_4),\ p(x_5) \}. \tag{2.3}$$

For an experiment consisting of one coin flip, there are two possible outcomes x_h and x_t, and the number of heads is either zero or one, which we can write as either $X = \{x_t, x_h\}$ or as $X = \{0, 1\}$, depending on how we wish to represent different values of the random variable X.

Within the Bayesian framework, coin bias is represented by the random variable Θ, which can adopt two possible values, $\theta_{0.1} = 0.1$ and $\theta_{0.9} = 0.9$. Thus, the probability distribution $p(\Theta)$ consists of just two probabilities, which we assume to be $p(\theta_{0.1}) = 0.75$ and $p(\theta_{0.9}) = 0.25$

$$
\begin{aligned}
p(\Theta) &= \{p(\theta_{0.1}), p(\theta_{0.9})\} & (2.4) \\
&= \{0.75, 0.25\}, & (2.5)
\end{aligned}
$$

where Θ is the upper case version of θ. Such a scenario might arise if we had a container in which 25% coins have a bias of 0.9, and 75% have a bias of 0.1, so that $p(\theta_{0.9})$ is the proportion of coins with a bias of 0.9, and $p(\theta_{0.1})$ is the proportion of coins with a bias of 0.1. If the bias of a coin is θ then the probability of a head x_h is, by definition, $p(x_h|\theta) = \theta$, and given that the probability of a head is the probability that $X = x_h$, the quantity $p(x_h|\theta)$, when written in full, is

$$p(X = x_h|\Theta = \theta) = \theta, \tag{2.6}$$

and is interpreted as: "The probability that the random variable $X = x_h$ (ie the probability that the coin lands heads up) given that the random variable Θ has the value θ, is equal to θ." This may appear more transparent if we set the coin bias to a specific value, like $\theta = 0.9$, so that Equation 2.6 becomes

$$p(X = x_h|\Theta = 0.9) = 0.9, \tag{2.7}$$

which is interpreted as: "The probability that a coin lands heads up given that it has a bias of 0.9 is 0.9."

2.2. The Rules of Probability

The founders of probability theory (Bayes, Bernoulli and Laplace) used informal notions of probability like those defined in Chapter 1, and regarded these rules as self-evidently true. Of course, the fact that the rules of probability can seem self-evident does not imply that the consequences that flow from these rules are obvious. In particular, even though Bayes' rule follows in a few lines of algebra from the rules of probability, no amount of staring at the rules themselves will make Bayes' rule obvious. Perhaps if it were more obvious, Bayes and others would not have had such trouble in discovering it, and we would not expend so much effort in understanding its subtleties. In the 20th century, these subtleties were explored in the modern incarnation of probability theory developed by Kolmogorov(1933)[24], Jeffreys(1939)[19], Cox(1946)[7], and Jaynes(2003)[18].

Below, we present three fundamental rules of probability, which are presented without proof (see Appendix C). Don't worry if these rules do not appear obvious at first, because they will be demonstrated several times in this chapter using geometric examples.

Independent Outcomes: If a set of individual outcomes are independent then the collective or *joint probability* of all the outcomes in that set is obtained by multiplying the probabilities of each of the individual outcomes together (see Appendix C). For example, consider a coin for which the probability of a head x_h is $p(x_h) = 0.9$, and the probability of a tail x_t is therefore $p(x_t) = (1-0.9) = 0.1$. If three coin flips yields a head followed by two tails then this defines the sequence (x_h, x_t, x_t). As none of the individual flip outcomes depend on the other two flip outcomes, all outcomes are independent of each other, so that the joint probability $p((x_h, x_t, x_t))$ of all three outcomes is obtained by multiplying the probabilities of each of the individual outcomes

$$
\begin{aligned}
p((x_h, x_t, x_t)) &= p(x_h) \times p(x_t) \times p(x_t) & (2.8) \\
&= 0.9 \times 0.1 \times 0.1 & (2.9) \\
&= 0.009. & (2.10)
\end{aligned}
$$

There is a subtle difference between calculating the probability of obtaining a particular sequence or *permutation* of three outcomes (as in this example), as opposed to the probability of obtaining a specific number of heads in *any order* or *combination* in three flip outcomes. Fortunately, this distinction is usually irrelevant for our purposes, but it is worth knowing about (see Appendix E).

The Sum Rule: As stated above, in our container of coins, some have a bias $\theta_{0.9} = 0.9$, and some have a bias $\theta_{0.1} = 0.1$. As there are only two possible values for the coin bias, the total probability $p(x_h)$ that a coin flip outcome yields a head is the sum of two joint probabilities:

1) $p(x_h, \theta_{0.9})$, the joint probability that a randomly chosen coin has a bias of 0.9 and that a head is observed, plus

2) $p(x_h, \theta_{0.1})$, the joint probability that a randomly chosen coin has a bias of 0.1 and that a head is observed.

In other words, the probability of observing a head is

$$p(x_h) \quad = \quad p(x_h, \theta_{0.9}) + p(x_h, \theta_{0.1}), \qquad (2.11)$$

where Equation 2.11 is the *sum rule*. In other words, the probability that $X = x_h$ is the sum of two joint probabilities: the joint probability that $[X = x_h$ and $\Theta = \theta_{0.9}]$ plus the joint probability that $[X = x_h$ and $\Theta = \theta_{0.1}]$.

The Product Rule: This is a subtle rule, which will be explained in detail later in this chapter. In essence, the product rule allows us to express a joint probability in terms of two other probabilities. For example, the joint probability $p(x_h, \theta_{0.9})$ can be expressed as

$$p(x_h, \theta_{0.9}) \quad = \quad p(\theta_{0.9}|x_h)p(x_h). \qquad (2.12)$$

In words, this means [the probability that a coin has a bias of $\theta_{0.9}$ and that a head is observed] equals [the probability that a coin has a bias of $\theta_{0.9}$ given that a head is observed] multiplied by [the overall probability of observing a head].

2.3. Joint Probability and Coin Flips

Now let's make our container of coins into a more concrete example by setting the number of coins to be $N = 100$, of which 25 have a bias of $\theta_{0.9} = 0.9$, and 75 have a bias of $\theta_{0.1} = 0.1$. Next, we are going to choose a coin, and flip it. Given that we pick a coin at random, we have chosen one of two possible bias values; in this case, let's assume it is $\theta_{0.9}$, and that we observe a head when we flip this coin. So, our pair of outcomes in this example is $(x_h, \theta_{0.9})$.

In this example, there are two types of event: the bias of a coin, and the outcome of flipping a coin. Because the coin bias can only adopt one of two values, and the outcome can also only adopt one of two values, there are a total of four possible pairs: $(x_h, \theta_{0.9})$, $(x_h, \theta_{0.1})$, $(x_t, \theta_{0.9})$, and $(x_t, \theta_{0.1})$.

Therefore, if we repeat the above procedure many times, and keep a record of each coin's bias and of each flip outcome, then our records depend on four joint probabilities:

1) $p(x_h, \theta_{0.9})$, the joint probability of choosing a coin with bias $\theta_{0.9}$ and that this coin lands heads up x_h.

2) $p(x_h, \theta_{0.1})$, the joint probability of choosing a coin with bias $\theta_{0.1}$ and that this coin lands heads up x_h.

3) $p(x_t, \theta_{0.9})$, the joint probability of choosing a coin with bias $\theta_{0.9}$ and that this coin lands tails up x_t.

4) $p(x_t, \theta_{0.1})$, the joint probability of choosing a coin with bias $\theta_{0.1}$ and that this coin lands tails up x_t.

In this experiment, each flip outcome depends on the bias of the particular coin chosen, so the two types of event are not independent.

	$\theta_{0.9}$	$\theta_{0.1}$	Sum
x_h	$p(x_h, \theta_{0.9}) = 0.225$	$p(x_h, \theta_{0.1}) = 0.075$	$p(x_h) = 0.300$
x_t	$p(x_t, \theta_{0.9}) = 0.025$	$p(x_t, \theta_{0.1}) = 0.675$	$p(x_t) = 0.700$
Sum	$p(\theta_{0.9}) = 0.250$	$p(\theta_{0.1}) = 0.750$	Total=1.00

Table 2.1: Joint and marginal probabilities of two correlated variables. Each of the 4 numbers in the centre of the table represents a joint probability. Each row sum is the probability of a head or a tail, and each column sum is the probability of a particular coin bias. Row and column sums are marginal probabilities, which must sum to one.

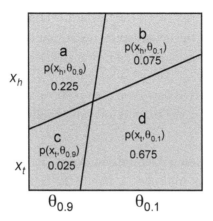

Figure 2.1: Joint probability as area. Given a coin with two possible bias values ($\theta_{0.1}$ or $\theta_{0.9}$), and an outcome with two possible values (heads=x_h or tails=x_t), there are four possible scenarios, $(\theta_{0.1}, x_t)$, $(\theta_{0.1}, x_h)$, $(\theta_{0.9}, x_t)$, $(\theta_{0.9}, x_h)$, each of which has a probability represented by the area of one quadrilateral. The area of all areas sum to one (not drawn to scale).

2.4. Probability As Geometric Area

Here, we show how joint probabilities behave just like areas, which allows the sum and product rules to be illustrated geometrically. This general idea is then extended to obtain expressions for conditional probabilities (eg likelihoods), which behave like ratios of areas.

The four joint probabilities defined in Section 2.3 represent the proportion of times each combination is observed. For example, the proportion of times that the coin chosen has a bias of $\theta_{0.9}$ and that a head x_h is observed turns out to be $p(x_h, \theta_{0.9}) = 0.225$, as in Table 2.1. Note that each row and column total is obtained by making use of the sum rule, and that these totals sum to one. These are traditionally written at the table margins, and are therefore called *marginal probabilities*. As each of the joint probabilities in Table 2.1 corresponds to a proportion, it can be represented by the area of a quadrilateral, as in Figure 2.1.

As the total area of the square is one, the proportion of the square occupied by each quadrilateral is numerically equal to the area of that quadrilateral. For example, if the total area of the square is $1m^2$ and if the area $a = 0.225m^2$ then the proportion of the square occupied by the area a is $0.225 (=0.225\text{m}^2/1\text{m}^2)$, which is also the joint probability $p(x_h, \theta_{0.9})$. Making the quadrilateral areas sum to one tallies with the fact that all 4 proportions and joint probabilities must sum to one, as shown in Table 2.1.

Before we can derive Bayes' rule, we need to consider the sum rule, posterior probability, the product rule and finally likelihood, in terms of areas.

The Sum Rule Using Geometric Area: In geometric terms, the quadrilaterals in the top 'row' in Figure 2.1 represent the areas that involve a head outcome, so the probability of a head is the proportion of area occupied by the top 'row' (ie $a+b$). Thus, the sum $a+b$ represents the probability $p(x_h)$ of a head

$$
\begin{aligned}
p(x_h) &= a + b & (2.13) \\
&= p(x_h, \theta_{0.9}) + p(x_h, \theta_{0.1}) & (2.14) \\
&= 0.3, & (2.15)
\end{aligned}
$$

where Equation 2.14 is the sum rule, summarised in Figure 2.2.

The Posterior Probability Using Geometric Area: Imagine we do not know the bias of a coin, and we want to estimate the probability that its bias is 0.9 (later we will compare this to the probability that its bias is 0.1, in order to decide which of these two biases it has). In case you have not recognised it, this is the posterior probability that a coin has a bias $\theta_{0.9}$.

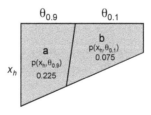

Figure 2.2: The posterior probability $p(\theta_{0.9}|x_h)$. Given that the outcome of a coin flip is a head x_h, the probability that the coin has a bias of $\theta_{0.9}$, is the ratio of areas $a/(a + b)$. This is an expanded view of part of Figure 2.1.

We begin by counting how often particular bias values and coin flip outcomes occur together. Accordingly, we choose a coin, make a note of its bias (marked on the coin), flip it, and make a note of the head/tail outcome. We repeat this 1,000 times, and keep a record of the 1,000 pairs of biases and head/tail outcomes. If there are 225/1,000 pairs that contain a coin with a bias of $\theta_{0.9}$ and a flip outcome that is a head then we estimate the joint probability of the bias as $\theta_{0.9}$ and a head to be 0.225. And if the number of heads is 300 then we estimate the probability of a head as 0.3.

Once all those bias markings have been removed from the coins, we choose one coin, flip it, and observe that it lands heads up. What is the probability that this coin has a bias $\theta_{0.9}$ given that a head x_h was observed? Looking back to our sample of 1,000 recorded results, it is the number of pairs that have a coin with a bias of $\theta_{0.9}$ and an outcome that is a head (225), expressed as a proportion of the total number of coins that landed heads up (300), which is 225/300=0.75.

When this is expressed in terms of areas, the joint probability $p(x_h, \theta_{0.9})$ that the coin chosen has a bias of $\theta_{0.9}$ and that coins chosen from our container landed heads up is the area a, whereas the overall probability $p(x_h)$ that a coin lands heads up is the area $(a + b)$ (see the sum rule above). Thus, the probability that a coin with bias $\theta_{0.9}$ was chosen given that a head x_h was observed is the area a expressed as a proportion of $a + b$

$$p(\theta_{0.9}|x_h) \quad = \quad a/(a + b), \tag{2.16}$$

where the area b corresponds to the joint probability $p(x_h, \theta_{0.1})$. We know that $a = p(x_h, \theta_{0.9})$, and (from Equation 2.13) that $a + b = p(x_h)$, so that Equation 2.16 evaluates to

$$p(\theta_{0.9}|x_h) \quad = \quad p(x_h, \theta_{0.9})/p(x_h) \tag{2.17}$$
$$= \quad 0.225/0.3 \tag{2.18}$$
$$= \quad 0.75, \tag{2.19}$$

which is a posterior or inverse probability (see Chapter 1).

The Product Rule Using Geometric Area: In order to find the posterior probability using Bayes' rule (instead of the joint probability, as above), we will need the product rule.

In geometric terms, the proportion of the top 'row' in Figure 2.1 occupied by the area a is $a/(a+b)$, and if this is multiplied by the total area of the top row $(a+b)$ then the result is the area a

$$a = \frac{a}{(a+b)} \times (a+b), \qquad (2.20)$$

where we already know that $a = p(x_h, \theta_{0.9})$, $a/(a+b) = p(\theta_{0.9}|x_h)$ and $(a+b) = p(x_h)$, which yields the product rule

$$p(x_h, \theta_{0.9}) \quad = \quad p(\theta_{0.9}|x_h)p(x_h). \qquad (2.21)$$

We can obtain the same result if we multiply both sides of Equation 2.17 by $p(x_h)$. If we substitute numbers into this ($p(\theta_{0.9}|x_h)$ from Equation 2.19 and $p(x_h)$ from Equation 2.15) then we obtain

$$p(x_h, \theta_{0.9}) \quad = \quad 0.75 \times (0.225 + 0.075) \qquad (2.22)$$

$$= \quad 0.225, \qquad (2.23)$$

as shown in Figure 2.1.

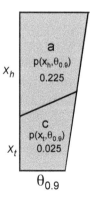

Figure 2.3: The likelihood $p(x_h|\theta_{0.9})$. Given a coin with a bias value of $\theta_{0.9}$, the probability of a head x_h is the ratio of areas $a/(a+c)$.

The Likelihood Using Geometric Area: In order to derive Bayes' rule, we will need the likelihood expressed as a ratio of areas.

As the likelihood is a conditional probability, we can use the same logic as was used to find the posterior probability, which is also a conditional probability. As before, we note that there are 225/1,000 pairs that contain a coin with a bias of $\theta_{0.9}$ and a flip outcome that is a head, so the joint probability of a bias $\theta_{0.9}$ and a head is estimated as 0.225. And if the total number of coins with a bias of $\theta_{0.9}$ is 250 then it follows that the probability of choosing a coin with a bias of $\theta_{0.9}$ is estimated as 0.25.

The likelihood $p(x_h|\theta_{0.9})$ is the probability that a head x_h was observed and that a coin with bias $\theta_{0.9}$ was chosen, expressed as a proportion of the total number of times that a coin with bias $\theta_{0.9}$ was chosen.

As before, we choose one coin, and observe that it has a bias of $\theta_{0.9}$ (the bias is marked on each coin, and has been left on for this experiment). What is the probability that this coin lands heads up given that it has a bias of $\theta_{0.9}$? By definition, the answer is 0.9, but we can compute this in a more roundabout way, and in a way that will prove useful below. Looking back to our sample of 1,000 recorded results, it is the number of pairs that had a coin with a bias of $\theta_{0.9}$ and an outcome that is a head (225), expressed as a proportion of the total number of coins chosen that have a bias of $\theta_{0.9}$ (250); it is 225/250=0.9.

When this is expressed in terms of areas, the probability that the coin chosen has a bias of $\theta_{0.9}$ and that a head was observed is the area a, whereas the proportion of times that a coin with bias $\theta_{0.9}$ was chosen is the area $(a + c)$ of the left 'column' in Figure 2.1, see Figure 2.3. Thus, the probability that a head was observed given that a coin with bias $\theta_{0.9}$ was chosen is the area a expressed as a proportion of $a + c$

$$p(x_h|\theta_{0.9}) \quad = \quad a/(a + c), \qquad (2.24)$$

where the area $c = p(x_t, \theta_{0.9})$, and $a + c$ represents the probability

$p(\theta_{0.9})$ that a coin with bias $\theta_{0.9}$ was chosen, as shown in Figure 2.3,

$$p(\theta_{0.9}) \quad = \quad a + c \qquad (2.25)$$

$$= \quad p(x_h, \theta_{0.9}) + p(x_t, \theta_{0.9}). \qquad (2.26)$$

Substituting this and $a = p(x_h, \theta_{0.9})$ into Equation 2.24 yields

$$p(x_h|\theta_{0.9}) \quad = \quad \frac{p(x_h, \theta_{0.9})}{p(\theta_{0.9})} \qquad (2.27)$$

$$= \quad 0.225/(0.225 + 0.025) \qquad (2.28)$$

$$= \quad 0.9, \qquad (2.29)$$

which is what we would expect given that this coin has a bias of 0.9. This is a forward probability, introduced in the previous chapter (p29).

Bayes' Rule Using Geometric Area: We already know from the product rule of Equation 2.21 that

$$p(x_h, \theta_{0.9}) \quad = \quad p(\theta_{0.9}|x_h)p(x_h). \qquad (2.30)$$

If we multiply Equation 2.27 by $p(\theta_{0.9})$ then we obtain another expression for this joint probability

$$p(x_h, \theta_{0.9}) \quad = \quad p(x_h|\theta_{0.9})p(\theta_{0.9}). \qquad (2.31)$$

We now have two expressions (Equations 2.30 and 2.31) for $p(x_h, \theta_{0.9})$, so we can write

$$p(\theta_{0.9}|x_h)p(x_h) \quad = \quad p(x_h|\theta_{0.9})p(\theta_{0.9}). \qquad (2.32)$$

Dividing both sides by $p(x_h)$ yields an equation for the posterior probability, but (unlike Equation 2.17) this is expressed in terms of the likelihood and the prior probability $(p(\theta_{0.9}))$ as Bayes' rule

$$p(\theta_{0.9}|x_h) \quad = \quad p(x_h|\theta_{0.9})p(\theta_{0.9})/p(x_h). \qquad (2.33)$$

Substituting numbers into Equation 2.33 yields

$$p(\theta_{0.9}|x_h) = (0.9 \times 0.25)/0.300 \qquad (2.34)$$
$$= 0.75, \qquad (2.35)$$

which agrees with Equation 2.19.

The line of reasoning that led to Equation 2.33 applies equally well to any value of θ, so we can substitute $\theta_{0.1}$ for $\theta_{0.9}$ in Equation 2.33 to obtain the posterior probability that the chosen coin has a bias of $\theta_{0.1}$ given that a head x_h was observed,

$$p(\theta_{0.1}|x_h) = p(x_h|\theta_{0.1})p(\theta_{0.1})/p(x_h). \qquad (2.36)$$
$$= (0.1 \times 0.75)/0.300 \qquad (2.37)$$
$$= 0.25. \qquad (2.38)$$

Thus, the posterior probability $p(\theta_{0.9}|x_h)$ that $\Theta = 0.9$ is 0.75, whereas the posterior probability $p(\theta_{0.1}|x_h)$ that $\Theta = 0.1$ is 0.25. The ratio of these two probabilities is the posterior odds (introduced on p14)

$$\frac{p(\theta_{0.9}|x_h)}{p(\theta_{0.1}|x_h)} = \frac{0.75}{0.25}. \qquad (2.39)$$

This tells us that, if a single coin flip yields a head then it is three times more probable that $\Theta = 0.9$ than it is that $\Theta = 0.1$. This is another example of model selection, introduced in Equation 1.24.

2.5. Bayes' Rule From Venn Diagrams

In this section, we again derive various quantities that allow us to prove Bayes' rule, but this time using Venn diagrams. This section is, to a large extent, a repeat of the previous section, so readers who need no further proof of Bayes' rule may wish to skip to the next chapter.

As in the previous example, we will define terms that allow us to evaluate the posterior probability $p(\theta_{0.9}|x_h)$ that the coin bias is $\theta_{0.9}$ given that a head x_h was observed. In Figure 2.4, the total area is one, and so the proportion of the total area occupied by the circle labelled A is equal to the probability $p(x_h)$ of observing a head, and encloses

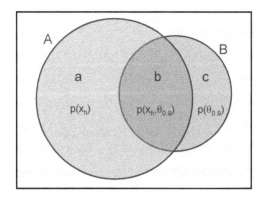

Figure 2.4: Bayes' rule from a Venn diagram, which has a total area of one. The area a of the disc A represents the probability $p(x_h)$ that a coin will land heads up, x_h. The area c of the disc B represents the probability $p(\theta_{0.9})$ that a coin has bias $\theta_{0.9} = 0.9$. The overlap between A and B is the area b, which represents the joint probability $p(x_h, \theta_{0.9})$ that a randomly chosen coin has bias $\theta_{0.9}$ and that a head x_h is observed.

an area a

$$p(x_h) = a. \tag{2.40}$$

In contrast, the proportion of the total area occupied the circle labelled B represents the probability $p(\theta_{0.9})$ that a coin with bias $\theta_{0.9} = 0.9$ is chosen, and encloses an area c

$$p(\theta_{0.9}) = c. \tag{2.41}$$

The area b enclosed by the intersection of A and B is the joint probability that a coin with bias $\theta_{0.9}$ is chosen, and that a head x_h is observed

$$p(x_h, \theta_{0.9}) = b. \tag{2.42}$$

Therefore, the probability that the coin has a bias of $\theta_{0.9}$ given that a head x_h is observed is the area b expressed as a proportion of a

$$p(\theta_{0.9}|x_h) \quad = \quad b/a. \tag{2.43}$$

Substituting Equations 2.42 and 2.40 in Equation 2.43, yields the posterior probability

$$p(\theta_{0.9}|x_h) \quad = \quad p(x_h, \theta_{0.9})/p(x_h). \tag{2.44}$$

Similarly, the probability of a head, given the coin has bias $\theta_{0.9}$ is

$$p(x_h|\theta_{0.9}) \quad = \quad b/c. \tag{2.45}$$

Substituting Equations 2.42 and 2.41 in 2.45, yields the likelihood

$$p(x_h|\theta_{0.9}) \quad = \quad p(x_h, \theta_{0.9})/p(\theta_{0.9}). \tag{2.46}$$

Bayes' Rule: If we multiply Equation 2.44 by $p(x_h)$, and Equation 2.46 by $p(\theta_{0.9})$ then we obtain, respectively,

$$p(x_h, \theta_{0.9}) \quad = \quad p(\theta_{0.9}|x_h)p(x_h) \tag{2.47}$$

$$p(x_h, \theta_{0.9}) \quad = \quad p(x_h|\theta_{0.9})p(\theta_{0.9}), \tag{2.48}$$

from which it follows that

$$p(\theta_{0.9}|x_h)p(x_h) \quad = \quad p(x_h|\theta_{0.9})p(\theta_{0.9}). \tag{2.49}$$

If we divide both sides by $p(x_h)$ then we obtain Bayes' rule

$$p(\theta_{0.9}|x_h) \quad = \quad p(x_h|\theta_{0.9})p(\theta_{0.9})/p(x_h). \tag{2.50}$$

2.6. Bayes' Rule and the Medical Test

In this section, we (yet) again derive various quantities that allow us to prove Bayes' rule, but this time using a graphical representation of a medical example. This is, to a large extent, a repeat of the previous two sections, so readers who need no further proof of Bayes' rule may wish to skip to the next chapter.

Each of the 100 squares in the small grid in Figure 2.5a represents 100 people, making a total population of 10,000 people. Within this population, just 1%, or a total of 100 people, have a disease. This 1% represents the background rate of the disease in the population. These 100 poor souls are represented by the square in the top right hand corner of Figure 2.5a, which is magnified in Figure 2.5b, where each square represents one of the 100 individuals who have the disease.

If a test for this disease is given, and 98 of the 100 diseased people have a positive result then the test for this disease has a *hit rate* of 98%, represented by 98 grey squares in Figure 2.5b. The *miss rate* of 2% is 100-(hit rate), so that two people who have the disease will have a negative result, represented by the 2 striped squares in Figure 2.5b.

Crucially, and this is a vital new piece of information, if the *false alarm* rate of the test is 3% then, out of the 9,900 people who do not have the disease, 297 of them will test positive, as represented by the three shaded squares (with 100 people per square) in Figure 2.5a.

Now, the probability that a person has the disease given that he has a positive test result is the (joint) probability that he has the disease and a positive test result, expressed as a proportion of the probability of a positive test result (the symbol +ve is short-hand for *positive*)

$$p(\text{disease}|+\text{ve test}) \quad = \quad p(+\text{ve test}, \text{disease})/p(+\text{ve test}). \quad (2.51)$$

Diagnosis Using the Joint Distribution: Let's take each term in Equation 2.51 in turn. The joint probability that a person has the disease and a positive test result is given by the total number of people who have the disease and who will also test positive (98) expressed as a proportion of the total population (10,000)

$$p(+\text{ve test}, \text{disease}) \quad \approx \quad 0.01, \quad\quad (2.52)$$

where the symbol $\approx$ means 'approximately equal to'. The overall probability of a positive test result is the sum of two probabilities: the probability $p(+\text{ve test}, \text{disease})$ that a person has the disease and a positive test result, plus the probability $p(+\text{ve test}, \text{no disease})$ that a

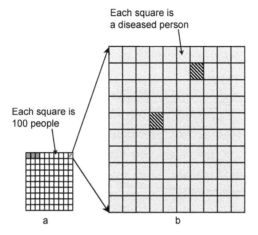

Figure 2.5: a) Each of the 100 squares represents 100 people, making a total of 10,000 people. The three black squares represent the 297 people who test positive, but have no disease. One square from (a) has been expanded to give the large grey grid (b), in which each of 100 squares represent an individual who has the disease. Of these 100 people, 98 will test positive, and two will test negative, as indicated by the two striped squares.

person does not have the disease and a positive test result

$$p(+\text{ve test}) = p(+\text{ve test}, \text{disease}) + p(+\text{ve test}, \text{no disease}), \quad (2.53)$$

which is another example of the sum rule. We know $p(+\text{ve test}, \text{disease})$ from Equation 2.52, but we have not calculated $p(+\text{ve test}, \text{no disease})$, the number of people with a positive test result who do not have the disease (297), expressed as a proportion of the total population (10,000)

$$p(+\text{ve test}, \text{no disease}) \quad = \quad 297/10,000 \qquad (2.54)$$

$$\approx \quad 0.03. \qquad (2.55)$$

Substituting Equations 2.55 and 2.52 into 2.53 yields

$$p(+\text{ve test}) \quad = \quad 98/10,000 + 297/10,000 \qquad (2.56)$$

$$\approx \quad 0.04. \qquad (2.57)$$

Substituting Equations 2.52 and 2.57 into 2.51 yields

$$p(\text{disease}|+\text{ve test}) \quad \approx \quad 0.01/0.04 \qquad (2.58)$$

$$= \quad 0.25. \qquad (2.59)$$

Thus, this test with a hit rate of 98% implies that a patient with a positive result has a 25% probability of having the disease. This surprising result is due to the low background rate $p(\text{disease})$ of the disease, and to the high false alarm rate $p(+\text{ve result}|\text{no disease})$ of the test. This is more apparent if we use the product rule (Equation 2.12)

$$p(+\text{ve test}, \text{disease}) \quad = \quad p(+\text{ve test} \mid \text{disease})\, p(\text{disease}), \quad (2.60)$$

to re-write the numerator of Equation 2.51 to obtain Bayes' rule

$$p(\text{disease}|+\text{ve test}) \quad = \quad \frac{p(+\text{ve test} \mid \text{disease})\, p(\text{disease})}{p(+\text{ve test})}. \quad (2.61)$$

Now we can see that even a large likelihood $p(+\text{ve test}|\text{disease})$ is consistent with a small posterior $p(\text{disease}|+\text{ve test})$ if the background rate (ie the prior $p(\text{disease})$) is small, or if the false alarm rate, which contributes to $p(+\text{ve test})$, is large.

Thus, given a background rate in the population of 1%, a test with a hit rate of 98% and a false alarm rate of 3% implies that a patient with a positive result has a 25% probability of having the disease.

Summary

The different geometric interpretations presented here have each been used to derive Bayes' rule. However, the subtleties of applying Bayes' rule to particular problems are far from obvious. These subtleties are explored in the remaining chapters of this book.

Chapter 3

Discrete Parameter Values

"I knew you came from Afghanistan. From long habit
the train of thoughts ran so swiftly through my mind,
that I arrived at the conclusion without being conscious
of intermediate steps. There were such steps, however. The
train of reasoning ran, 'Here is a gentleman of a medical
type, but with the air of a military man. Clearly an army
doctor, then. He has just come from the tropics, for his
face is dark, and that is not the natural tint of his skin,
for his wrists are fair. He has undergone hardship and
sickness, as his haggard face says clearly. His left arm has
been injured. He holds it in a stiff and unnatural manner.
Where in the tropics could an English army doctor have
seen much hardship and got his arm wounded? Clearly in
Afghanistan.' The whole train of thought did not occupy a
second. I then remarked that you came from Afghanistan,
and you were astonished."

Sherlock Holmes, from A Study in Scarlet. AC Doyle, 1887.

Introduction

The examples considered so far involve deciding between just two
possible alternatives, such as disease vs no disease, smallpox vs
chickenpox, or a coin bias of 0.1 vs 0.9. More usually, we want to
decide which one of several alternatives is most probably correct. In

this chapter, we discuss the application of Bayes' rule to more than two alternatives, but still to sets of separate, distinct possibilities, like a list of known diseases. Sets whose members are clearly separated from each other are called *discrete*, and the variables used to represent them are called *discrete variables*. The distribution of probability values of a discrete variable is called a *probability function*. The alternative to discrete variables are *continuous variables*, which have distributions called *probability density functions*, and these are considered in the next chapter.

3.1. Joint Probability Functions

Consider an array of boxes laid out in 4 horizontal rows, where each row represents a different set of symptoms, and 10 vertical columns, where each column represents a different disease, as in Table 3.1. As there are only 4 symptoms and only 10 diseases, they are represented as discrete variables.

The location of a box is given by its row number, which is represented by the discrete random variable

$$X \quad = \quad \{x_1, x_2, x_3, x_4\}, \tag{3.1}$$

and its column number, represented by the discrete random variable

$$\Theta \quad = \quad \{\theta_1, \theta_2, \theta_3, \theta_4, \theta_5, \theta_6, \theta_7, \theta_8, \theta_9, \theta_{10}\}, \tag{3.2}$$

(see Section 2.1 for a reminder of random variables). Given that there are $N_r = 4$ rows, X can adopt any value between $x_1 = 1$ and $x_4 = 4$. Similarly, given that there are $N_c = 10$ columns, Θ can adopt any value between $\theta_1 = 1$ and $\theta_{10} = 10$. For example, if we consider the box in row 3 and column 2 then the random variables X and Θ adopt the values $X = x_3$ and $\Theta = \theta_2$.

Sampling the Joint Probability Function

Just as we used $p(X)$ to represent an entire probability distribution in previous chapters, here, we use $p(X, \Theta)$ to represent an entire *joint probability distribution* or *joint probability function*. A joint probability function $p(X, \Theta)$ assigns a probability value to each box, where each box represents a particular combination of one disease θ and one set of symptoms x. This joint probability function can be visualised as a 3D histogram made up of square vertical columns, where the height of each column reflects the value $p(X, \Theta)$ of the probability assigned to the box below that column, as in Figure 3.1.

Now suppose that there are a total of $N = 200$ patients, and that some sort of random process (eg life) places each patient in one of the boxes, using the probabilities given by $p(X, \Theta)$. For example, it might be that the four values of X correspond to different age ranges, with higher values of Θ corresponding to diseases of older age ranges.

Boxes with high probability values usually end up with more patients than boxes with low probability values, as shown in Table 3.1 and Figure 3.2a. For example, if $p(x_3, \theta_9) = 0.04$ then we would expect there to be 8 (0.04×200) patients in the box located at (x_3, θ_9). However, because the process of assigning patients to boxes is probabilistic, there is no guarantee that the number of patients we would expect from each box's probability value is an exact match with the observed number of patients in that box.

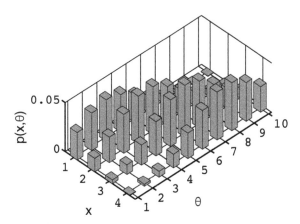

Figure 3.1: The discrete joint probability function $p(X, \Theta)$ defines a histogram where the height $p(x, \theta)$ of each column represents the probability that a patient has the particular combination of symptoms x and disease θ.

Thus, the spatial distribution of patients is said to represent a *sample* of the underlying joint probability function $p(X, \Theta)$. In fact, in order to see how well a given sample matches the joint probability function, we express the number of patients in each box as the proportion of the total number of patients in all boxes. For example, in a single sample of 200 patients, we may end up with 6 patients in box (x_3, θ_9). When expressed as a proportion of 200 patients, it is $6/200 = 0.03$, as opposed to the value $p(x_3, \theta_9) = 0.04$ of the joint probability function at (x_3, θ_9).

The process of assigning patients to each box is a bit like flipping a fair coin, inasmuch as we would not realistically expect to observe exactly 5 heads out of a sample size of 10 flips; we could reasonably expect the proportion of observed heads to be 0.6 (ie 6 heads), for example, or even 1.0 (ie 10 heads), very occasionally (see Section 4.7).

However, in real life, we never get to observe the bias of a coin directly; all we get is a number of heads and tails, which we can use to estimate the coin's bias. Similarly, we never get to observe a joint probability function $p(X, \Theta)$; at best, we get a sample from the joint probability function (eg Figure 3.2), from which we try to estimate joint probability function itself (actually, all we usually get is some summary information about the joint probability function, like the proportion of patients with each disease, see below). In other words, if patients get assigned to boxes according to the (unknown) probability values

X/Θ	θ_1	θ_2	θ_3	θ_4	θ_5	θ_6	θ_7	θ_8	θ_9	θ_{10}	Sum
x_4	0	0	1	0	3	5	10	7	7	4	37
x_3	0	1	1	10	16	11	12	7	8	5	71
x_2	3	5	8	9	14	10	3	3	0	0	55
x_1	8	9	9	5	4	1	1	0	0	0	37
Sum	11	15	19	24	37	27	26	17	15	9	200

Table 3.1: Joint distribution of symptoms x_r (in rows $r = 1-4$) and diseases θ_c (in columns $c = 1-10$). The number $n(x_r, \theta_c)$ in each cell represents the number of people with the symptoms x_r and disease θ_c. The totals at the table margins are proportional to two marginal distributions: the distribution of marginal likelihoods of symptoms (final column), and the prior distribution of diseases (bottom row).

specified by $p(X, \Theta)$ then, for any particular assignment of 200 patients, we could end up with 6, or 8 or even 200 patients in the box at (x_3, θ_9).

In order to simplify matters, we assume that the observed proportion of patients in every box is equal to the value of the joint probability function for that box (as we implicitly did in the medical example in Section 2.6). The reason for doing this is so that we can treat the proportion of patients in each box as if it were one value in the joint probability function. In practice, as explained above, all we have is the number of patients with each set of symptoms and the number of patients with each disease, so we are forced to make this assumption in any case. However, it is important to acknowledge that the numbers (strictly, proportions) we observe are estimates of some (unknown) underlying joint probability distribution, because we can then find ways to calculate how much confidence we can place in these estimates (see Section 4.7, p97).

There are two things worth noting here. First, as stated in the Ground Rules (p2), the sum of probabilities (eg symptom/disease

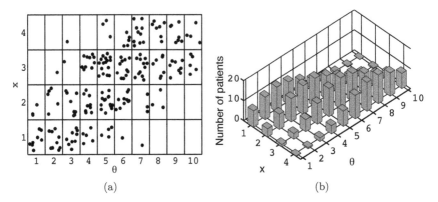

(a)   (b)

Figure 3.2: Visualising the joint probability function. **a)** The distribution of 200 patients. Here, the proportion of patients in each box is equal to the value of the joint probability function for that box. **b)** The height of each column equals the number of patients in the box underneath that column. This is exactly the same shape as Figure 3.1, but the total height of the vertical columns here is 200, whereas the heights in Figure 3.1 have been re-scaled so that the total height of the columns is equal to one.

combinations) must sum to one. This is because probabilities behave like proportions, and, just as we would expect all 40 proportions to sum to one, so we would expect all 40 probabilities to sum to one. This is a defining feature of a probability function, as in Figure 3.1. Second, even though the joint probability function $p(X, \Theta)$ is a discrete joint probability function because the random variables X and Θ are discrete, each of the 40 probability values it defines is represented by a continuous variable, with a value between zero and one.

Code Example 3.1: Plot Joint Distribution
File: Ch3Fig2a.py

```
# Make array of data, as in Table 3.1 (transposed here).
n = np.array([
[8,      9,      9,      5,      4,      1,      1,      0,      0,      0],
[3,      5,      8,      9,     14,     10,      3,      3,      0,      0],
[0,      1,      1,     10,     16,     11,     12,      7,      8,      5],
[0,      0,      1,      0,      3,      5,     10,      7,      7,      4]
])

fig = plt.figure()
ax = fig.add_subplot(111)
plt.ion() # Allow interactive plotting

for r in range(0, 4):
    for c in range(0, 10):
        ndots = n[r,c]
        for i in range(0,ndots):
            rr = 0.1+r+0.8*np.random.random()
            cc = 0.1+c+0.8*np.random.random()
            plt.scatter(cc,rr)

plt.ylim([0,4])
plt.xlim([0,10])
ax.yaxis.set_ticks([0,1,2,3,4])
plt.xlabel("$\\theta$")
plt.ylabel("x")
plt.grid(True)
plt.show()
# NB. Indexing in Python runs from 0 to N-1, but examples in book
# assumes indexing 1 to N. For consistency, the variable names
# used here assume 1 to N.
```

3.2. Patient Questions

If we know how many patients have each set of symptoms along with each disease then this tells us how many patients are in each box. Armed with this information, we can answer questions like these:

1. What is the joint probability $p(x_3, \theta_2)$ that a patient has the symptoms x_3 and the disease θ_2?

2. What is the probability $p(x_3)$ that a patient has symptoms x_3?

3. What is the probability $p(\theta_2)$ that a patient has the disease θ_2?

4. What is the conditional probability $p(x_3|\theta_2)$ that a patient has the symptoms x_3 given that he has the disease θ_2?

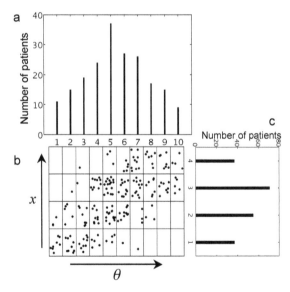

Figure 3.3: Figure (b) shows the distribution of 200 patients amongst 40 boxes, with $N_r = 4$ rows (symptoms) and $N_c = 10$ columns (diseases). The row number is indicated by the value of x, and the column number by θ. For example, the box at location (x_2, θ_8) contains 3 patients, each of which has symptoms x_2 and disease θ_8. Figure (a) is a histogram of the total number of patients in each column (ie for each disease θ), and (c) is a histogram of the total number of patients in each row (ie for each symptom x).

5. What is the conditional probability $p(\theta_2|x_3)$ that a patient has the disease θ_2 given that he has the symptoms x_3?

As we found in previous chapters, question 4 can be answered without any knowledge of the prevalence of different diseases, but this answer is almost useless, because it is based only on the available data (symptoms), and tells us only how probable the symptoms are given some putative disease (ie the likelihood). In contrast, the more interesting question 5 can only be answered by a doctor who combines the probability of the observed symptoms with prior knowledge of the prevalence of specific diseases in order to obtain the posterior probability of each disease. Because this tells us the probability of each putative disease, by extension, it also tells us the most probable disease for a given patient.

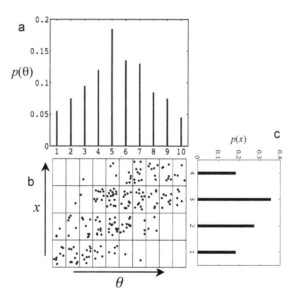

Figure 3.4: Joint and marginal probability distributions. As in Figure 3.3b, here (b) shows the joint distribution of 200 patients amongst a set of 40 boxes. However, the histograms in Figure 3.3a and Figure 3.3c which lie at the margins of Figure 3.3b have been normalised here so that the sum of column heights in each of the resultant *marginal distributions* is one. These marginal distributions are, (a) the prior probability distributions $p(\Theta)$ of diseases Θ, and, (c) the distribution of marginal likelihoods $p(X)$ of symptoms X.

```
# See Code Example 3.1 for definitions.
N = np.sum(n)                # Get total number of counts.
nx3theta2 = n[3,2]           # Number of counts in row=3 and col=2.
px3theta2 = nx3theta2/N      # Proportion in row x=3 and col theta=2.

print('Number␣of␣counts␣in␣n(3,2)␣is␣%d.' % nx3theta2)
print('Proportion␣of␣counts␣in␣p(3,2)␣is␣%.3f.' % px3theta2)

# Output: Number of counts in n(3,2) is 1.
#         Proportion of counts in p(3,2) is 0.005.
```

The Naming of Parts: Likelihood and Posterior: As discussed in Section 1.1 (p9), even though $p(x_3|\theta_2)$ and $p(\theta_2|x_3)$ are both conditional probabilities, and are therefore the same types of quantities from a logical viewpoint, we will treat them very differently. As in previous chapters, we will treat x_3 as some observed (ie known) data, and θ as a parameter whose value we wish to know. And, as a reminder, we refer to the conditional probability $p(x_3|\theta_2)$ as a likelihood, and the conditional probability $p(\theta_2|x_3)$ as a posterior probability.

Q1: What is the joint probability $p(x_3, \theta_2)$ that a patient has the symptoms x_3 and the disease θ_2?
In the above example, there are a total of $N = 200$ patients. The number of patients in the box located at (x_3, θ_2) is $n(x_3, \theta_2) = 1$. So, given our assumption that the proportion of patients with symptoms x_3 and the disease θ_2 equals the joint probability function at (x_3, θ_2),

$$p(x_3, \theta_2) \quad = \quad n(x_3, \theta_2)/N \qquad\qquad (3.3)$$

$$= \quad 1/200 \qquad\qquad (3.4)$$

$$= \quad 0.005. \qquad\qquad (3.5)$$

Q2: What is the probability $p(x_3)$ that a patient has the symptoms x_3?
In answering this question, we can illustrate the sum rule introduced in the previous chapter (also see Appendix C, p153). Specifically,

Question 2 can be answered by adding up patients in the third row, and then expressing this as a proportion of the total number of patients.

The total number of patients $n(x_3)$ in the third row can be obtained by summing all the patients in every box in the third row

$$n(x_3) \quad = \quad n(x_3, \theta_1) + n(x_3, \theta_2) + \cdots + n(x_3, \theta_{N_c}). \qquad (3.6)$$

Equation 3.6 can be written more succinctly if we make use of the summation notation

$$n(x_3) \quad = \quad \sum_{c=1}^{N_c} n(x_3, \theta_c) \qquad (3.7)$$

$$= \quad 71. \qquad (3.8)$$

For readers unfamiliar with this notation, the capital Greek letter $\sum$ (sigma) is interpreted as the sum of all terms to its right. The text at the base and top of the sigma letter tell us that the variable c is to be counted up from 1 to the total number N_c of columns. For each value of this 'counting' variable c, the term $n(x_3, \theta_c)$ adopts a new value, which is added to a running total in order to find the overall sum of all N_c quantities, laid out explicitly in Equation 3.6. Also see Appendix B.

Now, the number $n(x_3)$ corresponds to the height of the third row in the histogram shown in Figure 3.3c. This can be expressed as a proportion by dividing it by the total number N of patients, and, given our assumption that proportions equal probabilities,

$$p(x_3) \quad = \quad n(x_3)/N \qquad (3.9)$$

$$= \quad 71/200 \qquad (3.10)$$

$$= \quad 0.355, \qquad (3.11)$$

which corresponds to the height of the third column in the normalised histogram shown in Figure 3.4c.

Note that Equation 3.3 implies that each term in the summation on the right side of Equation 3.7 can be expressed as a probability

$$p(x_3, \theta_c) \quad = \quad n(x_3, \theta_c)/N. \qquad (3.12)$$

Code Example 3.3: Q2: Marginal Likelihoods
File: Ch3Q2.py

```
# See Code Example 3.1 for definitions.
N = np.sum(n)              # Get total number of counts.
row3 = n[2,:]              # Get row 3.
nx3=sum(row3)              # Number of counts in row=3.
px3 = nx3/N               # Proportion in row x=3.
X = np.sum(n,axis=1)/N    # Marignal distribution.

print ('Number of counts in row 3 is %d.' % nx3)
print('Proportion of counts in row 3 is %.3f.' % px3)
print('Marignal distribution p(X) ')
print(X)
# Output: Number of counts in row 3 is 71.
# Proportion of counts in row 3 is 0.355.
# Marignal distribution p(X) [ 0.185  0.275  0.355  0.185]
```

So, if we divide both sides of Equation 3.7 by N then we have

$$p(x_3) = \sum_{c=1}^{N_c} p(x_3, \theta_c). \qquad (3.13)$$

Thus, the probability $p(x_3)$ that a patient has symptoms x_3 is the joint probability that he has symptoms x_3 and disease θ_c, summed over all N_c diseases, as shown in Figure 3.4c.

The Sum Rule: More generally, the probability that a patient has the symptoms x is the sum of probabilities that a patient has each of the diseases and also has symptoms x

$$p(x) = \sum_{c=1}^{N_c} p(x, \theta_c). \qquad (3.14)$$

The process summarised in Equation 3.14 is called *marginalisation*, because it finds the distribution $p(X)$ along one margin of the joint distribution $p(X, \Theta)$. This particular marginal probability distribution is the distribution of marginal likelihoods of X. Here, the sum rule has been applied to the joint probability distribution $p(X, \Theta)$ for the row $X = x$, and involves summation over N_c values of the variable Θ in Equation 3.14. This illustrates a more general application of the sum rule (see Appendix C) than when it was applied to only two values of Θ in Equation 2.11.

When considered over all $N_r = 4$ rows, Equation 3.14 defines the *marginal probability distribution* $p(X)$ (Figure 3.4c and Table 3.1) of the joint distribution $p(X, \Theta)$ shown in Figure 3.4b. This marginal distribution consists of the *ordered set* of N_r probability values

$$
\begin{aligned}
p(X) &= [\, p(x_1), p(x_2), p(x_3), p(x_{N_r}) \,] & (3.15) \\
&= [\, 37,\ 55,\ 71,\ 37 \,]/200 \quad \text{(from Table 3.1)} & (3.16) \\
&= [\, 0.185,\ 0.275,\ 0.355,\ 0.185 \,], & (3.17)
\end{aligned}
$$

for which we use square brackets if the more conventional round brackets seem less than clear (as above).

Code Example 3.4: Q3: Prior Probability Distribution
File: Ch3Q3.py

```
# See Code Example 3.1 for definitions.
N = np.sum(n)              # Get total number of counts.
col2 = n[:,1]              # Get column 2.
ntheta2=sum(col2)          # Number of counts in col=2.
ptheta2 = ntheta2/N;       # Proportion in col=2.
ptheta = np.sum(n,axis=0)/N # Marignal distribution.

print('Number of counts in col 2 is %d.' % ntheta2)
print('Proportion of counts in col 2 is %.3f.' % ptheta2)
print('Marignal distribution p(THETA)')
print(ptheta)
# Output:
# Number of counts in col 2 is 15.
# Proportion of counts in col 2 is 0.075.
# Marignal distribution p(THETA)
# [0.055 0.075 0.095 0.12 0.185 0.135 0.13 0.085 0.075 0.045]
```

Q3: What's the probability $p(\theta_2)$ that a patient has disease θ_2?
To a large extent, this question is similar to the previous question, and can be answered using the same line of reasoning. It is equivalent to the question: What is the probability that a patient is in column θ_2? This can be answered by adding up all the patients in the second column, which we represent as $n(\theta_2)$, and then expressing this as a proportion of the total number of patients.

If we define the symbol N_r as the total number of rows, and $n(x_r, \theta_2)$ as the number of patients in row r of the second column, then the total number of patients $n(\theta_2)$ in the second column can be obtained by summing all the patients in every box in the second column

$$n(\theta_2) \;=\; \sum_{r=1}^{N_r} n(x_r, \theta_2). \tag{3.18}$$

$$= \;\; 9 + 5 + 1 + 0 \tag{3.19}$$

$$= \;\; 15. \tag{3.20}$$

Notice that $n(\theta_2)$ is the height of the second column in the histogram shown in Figure 3.3a. This can be expressed as a probability $p(\theta_2)$ by dividing it by the total number N of patients

$$p(\theta_2) \;=\; \sum_{r=1}^{N_r} n(x_r, \theta_2)/N, \tag{3.21}$$

$$= \;\; 15/200 \tag{3.22}$$

$$= \;\; 0.075, \tag{3.23}$$

where this probability corresponds to the height of the second column in the normalised histogram shown in Figure 3.4a.

Note that each term in the summation on the right side of Equation 3.21 is a proportion, and given our assumption that each observed proportion equals a probability,

$$p(x_r, \theta_2) \;=\; n(x_r, \theta_2)/N. \tag{3.24}$$

Substituting Equation 3.24 into 3.21 makes use of the sum rule again

$$p(\theta_2) \;=\; \sum_{r=1}^{N_r} p(x_r, \theta_2). \tag{3.25}$$

Thus, the probability $p(\theta_2)$ that a patient has the disease θ_2 is the sum of probabilities that a patient has the disease θ_2 and each of the $N_r = 4$ symptoms $x_1, \ldots, x_4$. When considered over all $N_c = 10$ columns, the marginal distribution (Figure 3.4a) of the joint distribution $p(X, \Theta)$ is

$$
\begin{aligned}
p(\Theta) &= [\, p(\theta_1), p(\theta_2), \ldots, p(\theta_{N_c}) \,] & (3.26) \\
&= [11, 15, 19, 24, 37, 27, 26, 17, 15, 9]/200 \quad \text{(from Table 3.1)} & (3.27) \\
&= [0.055, 0.075, 0.095, 0.120, 0.185, 0.135, 0.130, 0.085, 0.075, 0.045],
\end{aligned}
$$

shown in Figure 3.4b. This particular marginal probability distribution is the *prior probability distribution* of the random variable Θ.

Conditional Probability

Now we get to the more interesting questions, 4 and 5. These questions involve conditional probabilities, and are more interesting because they involve the types of quantities that actually get measured in the world.

As mentioned previously, we rarely have access to the joint distribution (or even to a sample of the joint distribution). If we did know the joint distribution then we could answer the questions above by simply finding the box in a particular row (or column) that has the most patients (and, indeed, we will demonstrate this obvious method below). Instead, we usually have access to the likelihood of each disease (which provides a forward probability) and, if we are lucky, to a prior distribution. Expressed in terms of medical diagnosis, this is because doctors studying, say, measles report the proportion of cases in which the patient has spots, which effectively provides the likelihood $p(\text{spots}|\text{measles})$. Similarly, doctors studying one of a range of other diseases (such as smallpox, chickenpox, flu, asthma) report the proportion of patients suffering from each of those diseases have spots. As these data constitute an estimate of the probability of having spots given each of a number of diseases, they effectively provide a likelihood for each disease. At the same time, different public health officials might compile statistics on the proportion of people in the general population who have each of those diseases. As this data set constitutes an estimate of the probability of having each disease, it effectively provides a prior probability distribution for the diseases under consideration. Crucially, when combined in accordance with Bayes' rule, the likelihoods and prior probabilities can be used to obtain the posterior probability that a specific patient with spots has each of a number of putative diseases.

Q4: What is the conditional probability $p(x_3|\theta_2)$ that a patient has the symptoms x_3 given that he has the disease θ_2? We begin by finding the total number $n(\theta_2)$ of patients in the second column θ_2, by adding up the patients in the $N_r = 4$ rows of that column

$$n(\theta_2) \quad = \quad \sum_{r=1}^{N_r} n(x_r, \theta_2) \qquad (3.28)$$

$$= \quad 15 \text{ patients.} \qquad (3.29)$$

Next, we count the number of patients in the box in the third row and second column, which is $n(x_3, \theta_2) = 1$ patient. Because there are a total of 15 patients in the second column, and one patient in the box at (x_3, θ_2), it follows that there is a one in 15 chance that a patient in the second column is the box in the third row x_3

$$p(x_3|\theta_2) \quad = \quad 1/15 \qquad (3.30)$$

$$= \quad 0.067. \qquad (3.31)$$

We can confirm this if we express $n(x_3, \theta_2)$ and $n(\theta_2)$ as probabilities

$$p(x_3|\theta_2) \quad = \quad \frac{n(x_3, \theta_2)/N}{n(\theta_2)/N} \qquad (3.32)$$

$$= \quad p(x_3, \theta_2)/p(\theta_2), \qquad (3.33)$$

$$= \quad 0.005/0.075 \qquad (3.34)$$

$$= \quad 0.067. \qquad (3.35)$$

This is the (forward) probability that a patient has symptoms x_3 given that he has disease θ_2, which is also the likelihood of the disease θ_2.

The Likelihood Function: Of course, this logic is not unique to the box at (x_3, θ_2), and applies equally well to any box in the third row. When considered over all possible values of Θ, the quantity $p(x_3|\Theta)$ is the probability of the observed data x_3 given different values of the parameter Θ, and defines the *likelihood function*,

$$
\begin{aligned}
p(x_3|\Theta) &= [p(x_3|\theta_1),\, p(x_3|\theta_2),\, \ldots,\, p(x_3|\theta_{N_c})] && (3.36) \\
&= [0.000, 0.067, 0.053, 0.417, 0.432, 0.407, 0.461, 0.412, 0.533, 0.556],
\end{aligned}
$$

where each term on the right side corresponds to the height of one column in the likelihood function of Figure 3.5. Using the same logic as in Equation 3.30, this graph was constructed from

$$
p(x_3|\Theta) = \left[\frac{n(x_3, \theta_1)}{n(\theta_1)},\, \frac{n(x_3, \theta_2)}{n(\theta_2)},\, \ldots,\, \frac{n(x_3, \theta_{N_c})}{n(\theta_{N_c})} \right]. \quad (3.37)
$$

Code Example 3.5: Q4: Likelihood Function
File: Ch3Q4.py

```python
# See Code Example 3.1 for definitions.
N = np.sum(n)                   # Get total number of counts.
ps = n/N                        # Make counts into proportions.
ntheta2 = np.sum(n[:,1]);       # Number of counts in col theta=2.
ptheta2 = ntheta2/N             # Proportion of counts in col theta=2.


px3ANDtheta2 = ps[2,1]
px3Gtheta2 = px3ANDtheta2/ptheta2

print('Prob␣x=3␣AND␣theta=2␣is␣%.3f.' % px3ANDtheta2)
print('Prob␣x=3␣given␣theta=2␣is␣%.3f.' % px3Gtheta2)


# Find likelihood function
nx3ANDTHETA = n[2,:]            # get theta values at x=3.
nTHETA = np.sum(n,axis=0)       # Marignal distribution.
px3GTHETA = nx3ANDTHETA/nTHETA

print( 'Likelihood␣function␣p(x3|THETA)␣=␣')
print(px3GTHETA)
# Output:
# Prob x=3 AND theta=2 is 0.005.
# Prob x=3 given theta=2 is 0.067.
# Likelihood function p(x3|THETA) =
# [0.000 0.067 0.053 0.417 0.432 0.407 0.462 0.412 0.533 0.556]
```

Arithmetic on Functions: Equation 3.37 is a convenient way of summarising all of the versions of Equation 3.32 for the possible different values of Θ. We can extend this method of summarisation to the sum rule, the product rule and Bayes' rule by giving a meaning to multiplication and division of functions. Given two functions with the same number of terms, such as

$$p(x_3|\Theta) = [p(x_3|\theta_1), p(x_3|\theta_2), \ldots, p(x_3|\theta_{10})] \qquad (3.38)$$
$$p(\Theta) = [p(\theta_1), p(\theta_2), \ldots, p(\theta_{10})], \qquad (3.39)$$

we define their product $p(x_3|\Theta)p(\Theta)$ as the sequence of products of individual terms, so that

$$p(x_3|\Theta)p(\Theta) = [p(x_3|\theta_1)p(\theta_1), \ p(x_3|\theta_2)p(\theta_2), \ \ldots, \ p(x_3|\theta_{10})p(\theta_{10})] \,.$$

Similarly, we define their quotient as the sequence of individual quotients. Now we can summarize all of the versions of Equation 3.33 for a particular value x_3 of X by

$$p(x_3|\Theta) = p(x_3, \Theta)/p(\Theta), \qquad (3.40)$$

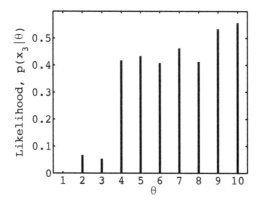

Figure 3.5: Likelihood function. The conditional probability $p(x_3|\theta)$ of symptoms x_3, given different values θ of Θ (diseases), defines a likelihood function $p(x_3|\Theta)$, which is the probability that a patient has symptoms defined by the third row x_3 given that he has the disease in each of the columns $\Theta = \{\theta_1, \ldots, \theta_{10}\}$. Note the values of a likelihood function do not necessarily add up to one (see p66).

and the product rule by

$$p(x_3, \Theta) \quad = \quad p(x_3|\Theta)p(\Theta). \tag{3.41}$$

That is, when written in full, Equation 3.41 means

$$p(x_3, \Theta) = [p(x_3, \theta_1), \ p(x_3, \theta_2), \ \ldots, \ p(x_3, \theta_{10})]$$

$$= [p(x_3|\theta_1)p(\theta_1), \ p(x_3|\theta_2)p(\theta_2), \ \ldots, \ p(x_3|\theta_{10})p(\theta_{10})],$$

which is just a summary of the product rules for all of the individual values of Θ. This type of function arithmetic only makes sense if, as in this example, the two functions have the same number of terms. For example, $p(X)/p(\Theta)$ would make no sense here because X has 4 elements, whereas Θ has 10 elements. We can extend our notation to allow multiplication and division of functions by numbers, for example,

$$p(\Theta)/p(x_3) \quad = \quad [p(\theta_1)/p(x_3), \ \ldots, \ p(\theta_{10})/p(x_3)], \tag{3.42}$$

where, again, we are just summarising a sequence of equations.

The Likelihood Function Is Not a Probability Function: By definition, the individual values of the probability function of a discrete variable sum to one. For example, when considered over all values of X (for a fixed value θ_2), the likelihoods defined by the probability function $p(X|\theta)$ sum to one

$$\sum_{r=1}^{N_r} p(x_r|\theta_2) = 1. \tag{3.43}$$

In terms of Figure 3.2, this amounts to expressing the number of patients with symptoms in each box in row x_r as a proportion of patients in the *same* column θ_2, and then summing these proportions across the rows of the same column, which therefore sum to one. In terms of symptoms and diseases, this amounts to adding up the proportions of patients with disease θ_2 who have different symptoms x_r, which therefore sum to one. This is akin to summing the proportion of girls and boys in the *same* school, a sum that must come to one.

In contrast, consider summing individual values from the likelihood function $p(x_3|\Theta)$, for a fixed value x_3,

$$\sum_{c=1}^{N_c} p(x_3|\theta_c).\tag{3.44}$$

This amounts to expressing the number of patients in each box in row x_3 as a proportion of the patients in column θ_c, and then summing these proportions across the *different* columns in row x_3. In terms of symptoms and diseases, this amounts to adding up the proportions of patients with each of the different diseases θ_c who have the same symptoms x_r. This is akin to summing the proportion of girls in *different* schools, and hoping that their sum will come to one. Whichever way this is expressed, there is no reason to suppose that these proportions should sum to one. This is why the likelihood function $p(x_3|\Theta)$ is not a probability function. Whilst this is of little significance most of the time, it is worth knowing.

Q5: What is the conditional probability $p(\theta_2|x_3)$ that a patient has the disease θ_2 given that he has symptoms x_3?
Using the same type of logic as in the previous section, we begin by finding the total number $n(x_3)$ of patients in the third row x_3, by adding up the patients in all of the $N_c = 10$ boxes in the third row

$$n(x_3) \quad = \quad \sum_{c=1}^{N_c} n(x_3, \theta_c) \tag{3.45}$$

$$= \quad 71 \text{ patients.} \tag{3.46}$$

We then count the number $n(x_3, \theta_2)$ of patients in the box that occupies the third row and second column $n(x_3, \theta_2) = 1$ patient. Since there are 71 patients in the third row, and one patient in the box located at (x_3, θ_2), there is a one in 71 chance that a patient in the third row comes from the box located at $n(x_3, \theta_2)$

$$
\begin{align}
p(\theta_2|x_3) &= n(x_3, \theta_2)/n(x_3) \tag{3.47} \\
&= 1/71 \tag{3.48} \\
&= 0.014. \tag{3.49}
\end{align}
$$

We obtain the same result if we divide $n(x_3, \theta_2)$ and $n(x_3)$ by N to yield proportions, and, if proportions equal probabilities then

$$
\begin{align}
p(\theta_2|x_3) &= \frac{n(x_3, \theta_2)/N}{n(x_3)/N} \tag{3.50} \\
&= p(x_3, \theta_2)/p(x_3), \tag{3.51}
\end{align}
$$

which is the (inverse) probability that a patient has the disease θ_2 given that he has the symptoms x_3. When considered over all possible values of Θ, the quantity $p(\Theta|x_3)$ is the probability of different values of the parameter Θ given the observed data x_3, and therefore defines

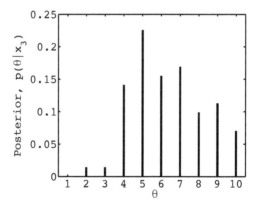

Figure 3.6: The posterior probability distribution. The conditional probability $p(\theta|x_3)$ of each parameter value θ, given the observed data x_3, defines a posterior probability distribution $p(\Theta|x_3)$. This is the probability that a patient has the disease in each of the columns $\theta_1, \ldots, \theta_{10}$ given that he has symptoms defined by the third row x_3 of Figure 3.4b.

the *posterior probability distribution*, shown in Figure 3.6,

$$p(\Theta|x_3) \quad = \quad p(x_3, \Theta)/p(x_3) \tag{3.52}$$

$$= \quad [p(x_3, \theta_1), p(x_3, \theta_2), \ldots, p(x_3, \theta_{N_c})]/p(x_3), \tag{3.53}$$

where we have implicitly assumed that the function $p(x_3, \Theta)$ can be divided by a single value $(p(x_3))$ by dividing each value of the function by $p(x_3)$. More generally, this applies to any particular symptom x, for which the posterior probability distribution is

$$p(\Theta|x) \quad = \quad p(x, \Theta)/p(x). \tag{3.54}$$

If we multiply both sides by $p(x)$ then we obtain an expression for a cross-section of the joint probability distribution $p(X, \Theta)$ at $X = x$ (see Sections 3.5, p76 and 6.4, p128)

$$p(x, \Theta) \quad = \quad p(\Theta|x)p(x), \tag{3.55}$$

which is another example of the product rule (eg Equation 3.41). This will be used with Equation 3.41 to derive Bayes' rule. (Readers content with the proof given in chapter 2 may wish to skip the next section).

Code Example 3.6: Q5: Posterior Probability
File: Ch3Q5.py

```python
# See Code Example 3.1 for definitions.
N = np.sum(n)                    # Get total number of counts.
ps = n/N                         # Make counts into proportions.
ntheta2 = np.sum(n[:,1])         # Number of counts in col theta=2.
ptheta2 = ntheta2/N              # Proportion of counts in col theta=2.
nx3 = np.sum(n[2,:])             # Number of counts in row x=3.

print('Number of counts in row x=3 is %d.' % nx3)

ptheta2Gx3 = n[2,1]/nx3
print('Prob x=3 given theta=2 is %.3f.' % ptheta2Gx3)

# Output:
# Number of counts in row x=3 is 71.
# Prob x=3 given theta=2 is 0.014.
```

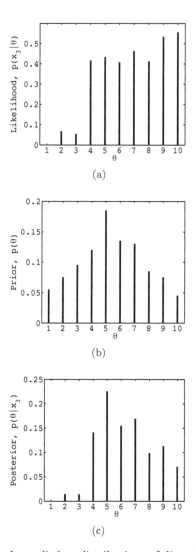

Figure 3.7: Bayes' rule applied to distributions of discrete variables. **a)** The likelihood function $p(x_3|\Theta)$ is the probability that a patient has symptoms x_3 given that he has each of the diseases $\theta_1 \ldots \theta_{10}$. **b)** The prior probability distribution $p(\Theta)$ is the probability that a patient has each of the diseases $\theta_1 \ldots \theta_{10}$. **c)** The posterior probability distribution $p(\Theta|x_3)$ is the probability that a patient has each of the diseases $\theta_1 \ldots \theta_{10}$ given that he has symptoms x_3. This involves multiplying corresponding column heights in the likelihood function (a) and the prior (b) distribution (see text).

Code Example 3.7: Plot Prior, Likelihood and Posterior Distributions
File: Ch3Fig7.py

```python
# See Code Example 3.1 for definitions.
N = np.sum(n)                          # Get total number of counts.
ps = n/N                               # Make counts into proportions.
nTHETA = np.sum(n,axis=0)

# Set graph parameters.
width = 0.2
theta_values = range(1,11)

# Find likelihood function.
nx3ANDTHETA = n[2,:]                   # Get theta values at x=3.
px3GTHETA = nx3ANDTHETA/nTHETA         # Likelihood p(x=3|theta).
# Plot likelihood function.
plt.figure("Figure 7a")
plt.bar(theta_values, px3GTHETA, width=width,align='center')
plt.xticks(theta_values, theta_values)
#  Add space to bar on left and right.
plt.xlim([min(theta_values)-0.5, max(theta_values)+0.5])
plt.xlabel("Outcome value")
plt.ylabel("Likelihood, $p(x_3|\\theta)$")

# Get prior distribution.
pTHETA = nTHETA/N                      # Prior distribution.
# Plot prior distribution.
plt.figure("Figure 7b")
plt.bar(theta_values, pTHETA, width=width,align='center')
plt.xticks(theta_values, theta_values)
plt.xlim([min(theta_values)-0.5, max(theta_values)+0.5])
plt.xlabel("Outcome value")
plt.ylabel("Prior, $p(\\theta)$")

# Find posterior distribution.
nx3 = np.sum(n[2,:])                   # Number of counts in row x=3.
px3 = nx3/N                            # Marginal probability p(x3)
pTHETAGx3 = px3GTHETA*pTHETA/px3       # Posterior distribution.
# Plot posterior distribution.
plt.figure("Figure 7c")
plt.bar(theta_values, pTHETAGx3, width=width,align='center')
plt.xticks(theta_values, theta_values)
plt.xlim([min(theta_values)-0.5, max(theta_values)+0.5])
plt.xlabel("Outcome value")
plt.ylabel("Posterior, $p(\\theta|x)$")
plt.show()
```

3.3. Deriving Bayes' Rule

Equations 3.41 and 3.55 give different expressions for a single cross-section of the joint distribution $p(X, \Theta)$ at $X = x$, so that

$$p(\Theta|x)p(x) \quad = \quad p(x|\Theta)p(\Theta). \tag{3.56}$$

Dividing both sides by the probability $p(x)$ yields Bayes' rule

$$p(\Theta|x) \quad = \quad p(x|\Theta)p(\Theta)/p(x), \tag{3.57}$$

where $p(\Theta|x)$ is the posterior probability distribution given a single observed value $X = x_3$, as described below.

Bayes' Rule for Probability Values: If we consider Bayes' rule in terms of individual probabilities then it can be written as

$$\text{posterior probability} \quad = \quad \frac{\text{likelihood} \times \text{prior probability}}{\text{marginal likelihood}}.$$

Given a value for the data, say $X = x_3$, and a specific value for the parameter Θ, such as, $\Theta = \theta_2$, Bayes' rule is written as

$$p(\theta_2|x_3) \quad = \quad p(x_3|\theta_2)p(\theta_2)/p(x_3) \tag{3.58}$$
$$= \quad (0.067 \times 0.075)/0.355 \tag{3.59}$$
$$= \quad 0.014. \tag{3.60}$$

Bayes' Rule for Probability Distributions: When Bayes' rule is considered over all possible values of the parameter Θ, only one term $p(x)$ is a scalar (a single value); of the remaining two terms, one is a likelihood function and the other is a probability distribution

$$\text{posterior distribution} \quad = \quad \frac{\text{likelihood function} \times \text{prior distribution}}{\text{marginal likelihood}},$$

which, for our example, translates to

$$p(\Theta|x_3) \quad = \quad p(x_3|\Theta)p(\Theta)/p(x_3). \tag{3.61}$$

If we multiply each of the N_c values of the likelihood within the likelihood function by each of the N_c corresponding values of the prior distribution then we obtain a value for each of the N_c posterior probabilities of the posterior distribution

$$
\begin{aligned}
p(\Theta|x_3) &= \frac{[p(x_3|\theta_1)p(\theta_1),\ p(x_3|\theta_2)p(\theta_2),\ \ldots,\ p(x_3|\theta_{N_c})p(\theta_{N_c})]}{p(x_3)} \\
&= [p(\theta_1|x_3),\ p(\theta_2|x_3),\ \ldots,\ p(\theta_{N_c}|x_3)]. \quad (3.62)
\end{aligned}
$$

This amounts to multiplying the heights of columns in the likelihood function (Figure 3.7a) by corresponding values of the prior distribution (Figure 3.7b), and dividing by the marginal likelihood $p(x_3)$, to obtain the posterior distribution shown in Figure 3.7c.

Evaluating the Marginal Likelihood: As discussed in the previous chapter, we usually do not need to know the value of the marginal likelihood $p(x_3)$. However, it is worth noting that, if we wanted the value of the term $p(x_3)$ then this can be found using marginalisation (introduced in Equation 3.14)

$$
p(x_3) = \sum_{c=1}^{N_c} p(x_3, \theta_c), \quad (3.63)
$$

where the product rule (Equation 3.41) allows us to re-write this as

$$
p(x_3) = \sum_{c=1}^{N_c} p(x_3|\theta_c)p(\theta_c). \quad (3.64)
$$

Note that this involves only those terms (the likelihood and the prior probabilities) which are also required to evaluate the numerator on the right side of Bayes' rule. Of course, this applies to each row in the joint distribution, and when considered over all values of X, this defines the distribution of marginal likelihoods $p(X)$ shown in Figure 3.4c

$$
p(X) = [p(x_1), p(x_2), \ldots, p(x_{N_r})]. \quad (3.65)
$$

Priors and Marginal Distributions: Notice that the prior distribution $p(\Theta)$ has been seen before in Equation 3.26 as a marginal distribution of the joint distribution $p(X, \Theta)$, in Figure 3.4a. Indeed, it is a general truth that the ideal prior distribution is the marginal of the joint distribution. Therefore, if we know the joint distribution then we can obtain the prior distribution using marginalisation (Equation 3.14). However, as we shall see, it is more often the case that we do not know the joint distribution, which is why we need Bayes' rule.

3.4. Using Bayes' Rule

Now let's see how Bayes' rule can be used to estimate the most probable value of the parameter Θ. Assume that we have observed some symptoms x_3, and that we want to estimate which of $N_c = 10$ possible diseases $\Theta = \{\theta_1, \theta_2, \ldots, \theta_{N_c}\}$ is most probable.

As explained in section 3.2 (p62), in practice, we usually do not have access to the joint probability distribution $p(X, \Theta)$ shown in Figures 3.1. However, we often know, or have good estimates of, crucial properties of this distribution, such as the prior distribution $p(\Theta)$, and the likelihood function $p(x_3|\Theta)$ (which can be used to find the marginal likelihood $p(x_3)$, as in Equation 3.64). These can then be used with Bayes' rule to obtain the posterior distribution of Equation 3.61. This is shown schematically in Figure 3.7, where Figure 3.7a is the likelihood function $p(x_3|\Theta)$, Figure 3.7b is the prior distribution $p(\Theta)$, and Figure 3.7c is the posterior distribution $p(\Theta|x_3)$, which is obtained by multiplying corresponding heights in Figure 3.7a and Figure 3.7b (all divided by $p(x_3)$). This allows us to identify the value of Θ that corresponds to the maximal value of $p(\Theta|x_3)$, which defines the *maximum a posteriori* (MAP) estimate of the true value of Θ. When considered over all values of Θ, the posterior distribution comprises $N_c = 10$ numbers

$$
\begin{aligned}
p(\Theta|x_3) &= [p(\theta_1|x_3), p(\theta_2|x_3), \ldots, p(\theta_{N_c}|x_3)] \quad\quad (3.66)\\
&= [0.00, 0.01, 0.01, 0.14, 0.22, 0.15, 0.17, 0.10, 0.11, 0.07],
\end{aligned}
$$

Code Example 3.8: Use Bayes' Rule to Estimate the Posterior Distribution
File: Ch3Section3.py

```
# See Code Example 3.1 for definitions.
N = np.sum(n)                        # Get total number of counts.
ps = n/N                             # Make counts into proportions.
nTHETA = np.sum(n,axis=0)            # Sum rows to get prior counts.


# Find likelihood function.
nx3ANDTHETA = n[2,:]                 # Get theta values at x=3.
px3GTHETA = nx3ANDTHETA/nTHETA       # Likelihood p(x=3|theta).
print("Likelihood function:")
print(px3GTHETA)


# Get prior distribution.
pTHETA = nTHETA/N                    # Prior distribution.
print("Prior distribution:")
print(pTHETA)


# Find posterior distribution.
nx3 = np.sum(n[2,:])                 # Number of counts in row x=3.
px3 = nx3/N                          # Marginal probability p(x3)
pTHETAGx3 = px3GTHETA*pTHETA/px3     # Bayes Rule=>Posterior distn.
print("Posterior distribution:")
print(pTHETAGx3)


# Output:
# Likelihood function:
# [0.000 0.067 0.053 0.417 0.432 0.407 0.462 0.412 0.533 0.556]
# Prior distribution:
# [0.055 0.075 0.095 0.120 0.185 0.135 0.130 0.085 0.075 0.045]
# Posterior distribution:
# [0.000 0.014 0.014 0.141 0.225 0.155 0.169 0.099 0.113 0.070]
```

where each number is the height of a column in Figure 3.7c. Clearly, the maximum value of the posterior distribution is $p(\theta_5|x_3)$, so that our MAP estimate of the true value of Θ is $\theta_{MAP} = \theta_5 = 5$.

Note the difference between the relatively flat likelihood function in Figure 3.7a and the 'peaky' posterior probability distribution in Figure 3.7c. Later, we shall see how more 'peaky' posterior probability distributions imply that we can place more confidence in the MAP estimate than in the MLE. But what is the true value of Θ? In one sense, it is not relevant to a Bayesian analysis, because we can never know the exact true value of any physical parameter. However, we can be reassured by the fact that a Bayesian estimate is the best that can be obtained (see Section 4.9).

3.5. Bayes' Rule and the Joint Distribution

We know from the derivation of Bayes' rule that the posterior probability distribution can be expressed in terms of a cross-section of the joint probability distribution $p(X, \Theta)$ at $X = x_3$

$$p(\Theta|x_3) \quad = \quad p(x_3, \Theta)/p(x_3) \tag{3.67}$$
$$= \quad p(x_3|\Theta)p(\Theta)/p(x_3). \tag{3.68}$$

However, if we had access to the full 2D joint distribution $p(X, \Theta)$ shown in Figure 3.3b then we would not need Bayes' rule at all. Given that $X = x_3$, we would simply count the patients in each box in row three, to find the one that contains the most patients. Recall that the number of patients in each box is related to the joint probability as

$$p(x_3, \Theta) \quad = \quad n(x_3, \Theta)/N. \tag{3.69}$$

If we find the value of Θ associated with largest value of $n(x_3, \Theta)$ then we have also found the value of Θ associated with largest value of the joint probability $p(x_3, \Theta)$. Here, the box with the largest number of patients (16) in row 3 is in column 5. Thus, the value of Θ associated with the largest value of $p(x_3, \Theta)$ is $\Theta = \theta_5$, and so the disease associated with the number 5 is our estimate of the true value of Θ.

But what about Bayes' rule? The reason that we did not make use of Bayes' rule in Equation 3.69 is because we had direct access to the joint probability distribution $p(x, \Theta)$, which made it easy to find the box in row 3 that had the most patients. But the product rule ensures that a cross-section of the joint probability distribution at $X = x_3$ is

$$p(x_3, \Theta) \quad = \quad p(\Theta|x_3)p(x_3). \tag{3.70}$$

It is important to note that this equation says the cross-section of the joint distribution $p(x_3, \Theta)$ is proportional to the posterior $p(\Theta|x_3)$. This matters because it means that the value of $p(x_3)$ does not affect the overall 'shape' of the posterior $p(\Theta|x_3)$ (see Section 4.5, p91), which ensures that $p(x_3, \Theta)$ and $p(\Theta|x_3)$ have peaks at the same value of Θ.

Consequently, the value θ_{MAP} below the peak in $p(\Theta|x_3)$ is the same as the value below the peak in the cross-section $p(x_3, \Theta)$. Thus, as we do not usually know the joint distribution $p(X, \Theta)$, we use Bayes' rule, because both approaches yield exactly the same estimate of the true value of the parameter Θ.

Summary

We have scrutinised a particular joint distribution, which was used to visualise its two marginal distributions, and then used to find the likelihoods and posterior probabilities of specific events. This joint distribution was defined to be the spatial distribution of patients in an array of boxes, which was also interpreted in terms of symptoms and diseases. Accordingly, each of the likelihoods considered above can be interpreted as the probability of symptoms given some disease, and each of the posterior probabilities can be interpreted as the probabilities of a disease given certain symptoms. We discovered that the most probable disease associated with an observed set of symptoms can be obtained from the joint probability distribution, which is usually unavailable. However, the most probable disease associated with an observed set of symptoms can also be obtained from the posterior probability distribution, which can be found using Bayes' rule.

Chapter 4

Continuous Parameter Values

> If we have no information relevant to the actual value of the parameter, the [prior] probability must be chosen so as to express the fact that we have none.
> H. Jeffreys, Theory of Probability, 1939.

Introduction

In this chapter, we explore Bayes' rule for continuous variables. The values of a continuous variable are like densely packed points on a line, where each point corresponds to the value of a *real number*. The main advantage of working with continuous values is that we can usually describe a probability distribution with an equation, and because an equation is defined in terms of a few key parameters, it is said to provide a *parametric* description of a probability distribution.

We use the example of coin bias here because it is easy to understand, but this type of analysis could apply to any situation in which a (discrete or continuous) variable x depends on the value of a continuous parameter θ. For example, the continuous parameter θ could represent temperature, and x could represent the (discrete) number of ice creams sold, or the (continuous) amount of water that evaporates from a lake.

4.1. A Continuous Likelihood Function

Consider a coin with a bias of $\theta_{true} = 0.6$ (ie it lands heads-up 60% of the time) that is flipped a total of $N = 10$ times. Suppose that each of the first 7 outcomes is a head x_h, and each of the final 3 outcomes is a tail x_t. The first 7 outcomes define the sequence or *permutation*

$$\mathbf{x}_h = (x_h, x_h, x_h, x_h, x_h, x_h, x_h). \tag{4.1}$$

The probability of a head given a putative value θ of the coin's bias is $p(x_h|\theta) = \theta$, and, since the outcome of each coin flip is independent of all other coin flips, the conditional probability of the permutation $\mathbf{x}_h$ is

$$
\begin{aligned}
p(\mathbf{x}_h|\theta) &= p((x_h, x_h, x_h, x_h, x_h, x_h, x_h)|\theta) & (4.2) \\
&= p(x_h|\theta)^7 & (4.3) \\
&= \theta^7. & (4.4)
\end{aligned}
$$

The final 3 flips define a permutation of tails $\mathbf{x}_t = (x_t, x_t, x_t)$, where the probability of each tail x_t is $p(x_t|\theta) = (1 - \theta)$, so the conditional probability of $\mathbf{x}_t$, given θ, is

$$
\begin{aligned}
p(\mathbf{x}_t|\theta) &= p((x_t, x_t, x_t)|\theta) & (4.5) \\
&= p(x_t|\theta)^3 & (4.6) \\
&= (1 - \theta)^3. & (4.7)
\end{aligned}
$$

We can define the permutation of 7 heads followed by 3 tails as the concatenation of $\mathbf{x}_h$ and $\mathbf{x}_t$, which yields the permutation $\mathbf{x} = (x_h, x_h, x_h, x_h, x_h, x_h, x_t, x_t, x_t)$. As $\mathbf{x}_h$ and $\mathbf{x}_t$ represent mutually independent sets of events, the probability of $\mathbf{x}$ is

$$
\begin{aligned}
p(\mathbf{x}|\theta) &= p(\mathbf{x}_h|\theta)p(\mathbf{x}_t|\theta) & (4.8) \\
&= \theta^7(1 - \theta)^3, & (4.9)
\end{aligned}
$$

where Equation 4.9 is a type of *binomial equation* (see Appendix E).

```python
# Define vector of values for theta.
THETA = np.linspace(0.0, 1.0,100)
# Find likelihood function from Eq 4.9.
pxGTHETA = THETA**7 * (1.0-THETA)**3
# Plot likelihood function, Figure 4.1.
plt.figure("Figure 4.1")
plt.plot(THETA,pxGTHETA)
plt.xlabel("Bias, $\\theta$")
plt.ylabel("Likelihood function $p(x|\\theta)$")
plt.show()
```

Finding the Maximum Likelihood Estimate: When considered over all values Θ of θ this defines the likelihood function

$$p(\mathbf{x}|\Theta) = [p(\mathbf{x}|\theta_1), p(\mathbf{x}|\theta_2), \ldots] \qquad (4.10)$$
$$= [\theta_1^7(1-\theta_1)^3, \theta_2^7(1-\theta_2)^3, \ldots] \qquad (4.11)$$
$$= \Theta^7(1-\Theta)^3. \qquad (4.12)$$

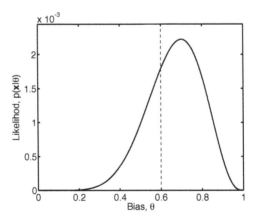

Figure 4.1: Likelihood function. A graph of the likelihood of θ, the probability $p(\mathbf{x}|\Theta)$ of $x = 7$ heads out of $N = 10$ coin flips, under the assumption that the coin bias Θ adopts a range of values. The peak of the likelihood function $p(\mathbf{x}|\Theta)$ is at $\theta = 0.7$, which is the maximum likelihood estimate (MLE) of the true value of Θ. The true value $\theta_{true} = 0.6$ is indicated by the vertical dashed line. Note the y-axis is multiplied by $10^{-3} = 1/1000$.

We can plot the likelihood function for the data $\mathbf{x}$ by evaluating Equation 4.12 for different values of θ, so each term on the right side corresponds to the height of one point in the likelihood function of Figure 4.1. We can then observe that, when $\theta = 0.7$, the likelihood function reaches its maximum value of

$$p(\mathbf{x}|\theta) \quad = \quad \theta^7(1-\theta)^3 = 0.0022. \tag{4.13}$$

The value of θ that yields the highest value of the likelihood function is the maximum likelihood estimate (MLE) of the true value $\theta_{true} = 0.6$; that is, the maximum likelihood estimate is $\theta_{MLE} = 0.7$. More generally, it can be shown that the MLE of the true value of the coin's bias is given by the proportion of observed heads.

In the process of plotting Figure 4.1, we have effectively 'tried out' different values of θ in Equation 4.13 to see which value is most consistent with 7 out of 10 heads; that is, the value of θ that makes 7/10 heads most probable. The result represents our best estimate θ_{MLE} of the coin's true bias θ_{true} based on the observed data (but disregarding any information based on previous experience). Note that a different number of heads would yield a different likelihood function (eg see Figure 4.7), with a different value for the MLE.

The probability of the data? At first, this way of thinking about the data seems odd. It just sounds wrong to speak of the probability of the data, which are the things we have already observed; so why would we care how probable they are? In fact, we do not care about the probability of the data *per se*, but we do care how probable those data are in the context of parameters we wish to estimate; that is, in the context of our binomial model (Equation 4.9) of how the data were generated. Specifically, within our model, we care about the value of θ that would make the observed data most probable (ie the value of θ that is most consistent with the data).

Because we now have an equation that describes how the likelihood changes with respect to θ, we can find the MLE without having to evaluate the likelihood function at many (often many thousands) of values. In essence, once we have an equation for the likelihood function, we know that the slope of this function is zero at the MLE, as can be seen from Figure 4.1. Using calculus, we can find an expression for the slope, and if we then solve for the value of θ that makes this expression equal to zero then we have found the MLE. The details of this type of analysis are shown in the context of finding the maximum value of a posterior distribution in Section 4.6 (p93), but the same principles are usually used to find the MLE.

Pdfs and pfs

There is a subtle change in terminology that distinguishes between the probability distributions associated with discrete and continuous variables. With the proviso that the area under a probability distribution is one, the distribution of probabilities of a discrete variable is called a *probability function* (pf), or a *joint probability function* if there is more than one variable, whereas the distribution of probabilities of a continuous variable is called a *probability density function* (pdf), or a *joint probability density function* if there is more than one variable (see Appendix D, p157). Thus, the distribution of prior and posterior probabilities defines a pf if they are discrete and a pdf if they are continuous. In the case of a pf, the sum of all heights on the distribution

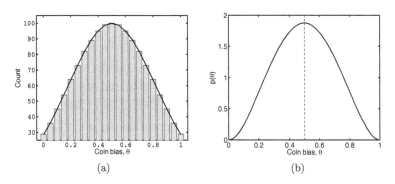

(a)   (b)

Figure 4.2: A prior distribution for coin bias. **a)** Based on 1,432 coins, the distribution of estimated biases is plotted here as a histogram with 21 columns. **b)** A binomial function fitted to the normalised histogram approximates the prior probability density function (pdf) of coin bias.

is one, and, for a pdf, the total area under the distribution is one. In contrast, the corresponding sum for a distribution of likelihood values is not necessarily one (see p66), so it is not a pf nor a pdf, and is often referred to as an *un-normalised distribution*. For simplicity, we use the term *distribution* to refer to pfs, pdfs, and likelihood functions because it is usually clear which type of distribution is being referred to.

Code Example 4.2: A Binomial Prior
File: Ch4Fig2b.py

```
# Define vector of values for theta.
THETA = np.linspace(0.0, 1.0,100)
# Find prior distribution from Eq 4.14.
pTHETA = THETA**2 * (1.0-THETA)**2
# Plot prior distribution, Figure 4.2b.
plt.figure("Figure 4.2b")
plt.plot(THETA,pTHETA)
plt.xlabel("Bias, $\\theta$")
plt.ylabel("Prior probability density $p(\\theta)$")
plt.show()
```

4.2. A Binomial Prior

Let's suppose that we have a vast wealth of experience in estimating coin biases, and that we have kept a record of the estimated bias of every coin we have ever encountered. We can plot the frequency with which each bias occurs as a histogram, as in Figure 4.2a. This histogram represents all the knowledge implicit in our previous experience of coin biases. In effect, it provides a hint about the probable value of a coin's bias before we have flipped that coin once. Notice that the height of the histogram has a maximum value at $\theta = 0.5$, and this is the mean bias of all of the coins. Also, the width of the histogram provides a clue regarding the certainty of our prior knowledge of a coin's bias. For example, if the histogram were very narrow then we would be reasonably certain that any new coin we choose will have a bias close to $\theta = 0.5$. Conversely, if the histogram were quite broad (but still centred on $\theta = 0.5$) then our best guess at a new coin's bias would still be $\theta = 0.5$, but we could be much less certain that its value will be close to 0.5.

If we normalise this histogram so that it has an area of one (ie the sum of all probabilities is one), and fit a curve to it, then this curve is a good approximation to a probability density function (see Appendix D). As this particular probability density function refers to prior probabilities, it is called a *prior probability density function*. Purely for convenience, we define the prior probability density function with the binomial equation (see Appendix E)

$$p(\Theta) \quad = \quad \Theta^2 (1 - \Theta)^2. \tag{4.14}$$

If we substitute values of Θ between zero and one into Equation 4.14 then we obtain the graph of the prior shown in Figure 4.2b (also 4.3a).

Note that we only ever get to see a finite sample of data (eg coin biases, coin flip outcomes), but we do not get to see the continuous distribution from which this sample is drawn. Thus, it is not possible to observe a continuous distribution, but we often find it convenient to act as if such a distribution exists, and that it can be described with an equation. This equation then acts as a model for the observed data, so that the observed data get treated as an approximation to the idealised distribution of the model.

4.3. The Posterior

We now have an equation (4.12) for the likelihood function $p(\mathbf{x}|\Theta)$, and an equation (4.14) for the prior probability density function $p(\Theta)$. Using Bayes' rule, these can be combined to obtain the posterior probability density function $p(\Theta|\mathbf{x})$, and this can be used to choose the best value of the continuous parameter Θ.

Recall that the posterior probability of each of two diseases was obtained by weighting the likelihood of each disease by its prior probability. This amounts to multiplying each value of the likelihood function $p(\mathbf{x}|\Theta)$ (Figure 4.3b) by the corresponding value of the prior probability density function $p(\Theta)$ (Figure 4.3a) to obtain the *posterior probability density function* $p(\Theta|\mathbf{x})$ (Figure 4.3c).

Rather than finding the posterior distribution for each of several discrete values of Θ (as in the previous chapter), we can use Bayes'

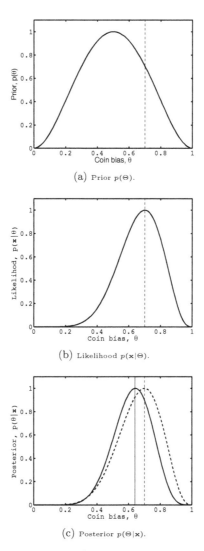

(a) Prior $p(\Theta)$.

(b) Likelihood $p(\mathbf{x}|\Theta)$.

(c) Posterior $p(\Theta|\mathbf{x})$.

Figure 4.3: Bayes' rule in action. **a**) A prior probability density function $p(\Theta)$ for coin bias θ. **b**) The likelihood function $p(\mathbf{x}|\Theta)$ for $x = 7$ heads. The peak defines the maximum likelihood estimate of θ_{true} as $\theta_{MLE} = 0.7$ (vertical line). **c**) Multiplication of corresponding heights on the prior pdf $p(\Theta)$ and the likelihood function $p(\mathbf{x}|\Theta)$ yields a scaled version of the posterior pdf $p(\Theta|\mathbf{x})$ (solid curve). Its peak defines the maximum a posteriori estimate of θ_{true} as $\theta_{MAP} = 0.64$. The dashed curve is the likelihood function (from b). Graphs re-scaled to have a maximum value of one.

rule to derive an equation for the continuous posterior pdf

$$p(\Theta|\mathbf{x}) = p(\mathbf{x}|\Theta)p(\Theta)/p(\mathbf{x}) \qquad (4.15)$$

$$= \Theta^7(1-\Theta)^3 \times \Theta^2(1-\Theta)^2 \qquad (4.16)$$

$$= \Theta^9(1-\Theta)^5, \qquad (4.17)$$

where we have disregarded the constant $p(\mathbf{x})$ in Equations 4.16 and 4.17 (see Section 4.5). By substituting individual values of Θ between zero and one, this equation can be used to plot the posterior probability density function shown in Figure 4.3c, where the peak corresponds to the maximum a posteriori (MAP) estimate of θ_{true}. In this instance, we find that $\theta_{MAP} = 0.64$, so this is our MAP estimate of the unknown value θ_{true} based on 10 coin flips (compared to $\theta_{MLE} = 0.7$). More importantly, once we have an equation for the posterior probability density function, we no longer need graphs like Figure 4.3c (even though they are useful for visualising a pdf), because we can use calculus to find the MAP analytically (see Section 4.6).

Code Example 4.3: Plot Posterior Distribution
File: Ch4Fig3c.py

```
# Define vector of values for theta.
THETA = np.linspace(0.0, 1.0,100)

# Find posterior distribution, Equation 4.17.
pTHETAGx = THETA**9 * (1.0-THETA)**5
# Set max value to 1, for graph.
pTHETAGx = pTHETAGx/max(pTHETAGx)
# Plot posterior distribution, Figure 4.3c.
plt.figure("Figure 4.3c")
plt.plot(THETA,pTHETAGx)
plt.xlabel("Coin bias, $\\theta$")
plt.ylabel("Posterior probability density $p(\\theta|x)$")

# Find likelihood function from Eq 4.9.
pxGTHETA = THETA**7 * (1.0-THETA)**3
# Set max value to 1, for graph.
pxGTHETA = pxGTHETA/max(pxGTHETA)
# Plot likelihood function, for comparison.
plt.plot(THETA,pxGTHETA,'r--')
plt.ylim([0, 1.1])
plt.show()
```

4.4. A Rational Basis For Bias

For a coin which lands heads up with probability θ_{true}, we use the observed proportion x of heads to estimate θ_{true}. But all measurement devices are imperfect (eg we can mis-count the number of heads), so the measured estimate x of the true proportion of heads x_{true} is noisy. Additionally, the proportion x_{true} of heads is usually probabilistically related to θ_{true} (eg the proportion x_{true} of heads varies randomly around θ_{true}). So there are at least two sources of uncertainty: uncertainty in the measured value x, and uncertainty in the relation between x and the parameter Θ (see Figure 1.12, p28).

These sources of uncertainty translate to a corresponding uncertainty in the value of Θ, which defines a likelihood function $p(x|\Theta)$. However, if we know the underlying (ie prior) distribution $p(\Theta)$ of values of Θ in the world then we can use this as a guide to reduce the uncertainty in Θ. In essence, this is what Bayes' rule does. So, if we seek the most probable value θ_{MAP} of Θ then, given a measurement x, Bayes' rule tells us how to 'adjust' the estimated value θ_{MLE} of θ_{true}, so that (on average), the adjusted value θ_{MAP} is more accurate than θ_{MLE}.

Thus, the available information x is incomplete, either because our measurement x adds noise to x_{true}, or because even a noise-free measurement $x = x_{true}$ is probabilistically related to the value θ_{true} of the parameter Θ. In both cases, Bayes' rule provides a rational basis for imposing a particular choice of values (the prior, which may appear to be biased) on estimated parameter values, to arrive at a value that represents our estimate of the most probable state of the physical world.

4.5. The Uniform Prior

What if we have no previous experience of an unknown parameter like coin bias, and therefore no reason to prefer any particular coin bias values over any others? In such cases, we have two options.

First, we could simply rely on the observed data (coin flip outcomes) in the form of a likelihood function, which implicitly acknowledges our lack of knowledge or previous experience of the value of a parameter. Note that this implicitly uses a uniform prior for the parameter.

Second, we could make our ignorance explicit in the form of a particular prior probability density function. As we do not know how much weight to assign to each likelihood value, its seems 'fair' to weight them all equally, which defines a *uniform prior probability density function.*

In practice, these two options have exactly the same outcome, as can be seen by comparing the likelihood function in Figure 4.4b with the (identical) posterior probability density function in Figure 4.4c. So why bother with a uniform prior? Because the general strategy of making our ignorance explicit by replacing the implicit uniform prior with an explicit prior forces us to be clear about the nature of our ignorance. The reason this matters is because sometimes the 'fairest' prior is not a uniform prior. Such non-uniform priors do not occur despite our ignorance, but because of it. In essence, the Bayesian framework forces us to express our ignorance regarding the value of a parameter in a principled manner. The result is a 'fair' or *reference prior* that is not necessarily uniform. In fact, the reference prior for coin bias is not uniform, but we will assume that it is for the present, and we will explore the use of reference priors in Section 4.8, p99.

If we know nothing about the underlying distribution of Θ then it seems natural to assign a uniform prior probability density function $p(\Theta) = c$, where c is a constant, as shown in Figure 4.4a. If the range of plausible values of θ is between θ_{min} and θ_{max} then the area under the prior probability density function is $c\,(\theta_{max} - \theta_{min})$. As Θ must adopt one value, the sum of all probabilities of $p(\Theta)$ must be one, so that $c\,(\theta_{max} - \theta_{min}) = 1$, and therefore $c = 1/(\theta_{max} - \theta_{min})$. In this particular example, $\theta_{min} = 0$ and $\theta_{max} = 1$, so the range of the prior is one, which yields $c = 1$. So, each likelihood value receives the same weighting from the prior probability density function, which, in this case, is a weighting of one. More generally, if c is any constant value then the posterior probability density function has exactly the same shape as the likelihood function, shown in Figure 4.4c.

As noted above, a uniform prior probability density function yields a posterior probability density function that has the same shape as the likelihood function. This, in turn, means that the location of the peak

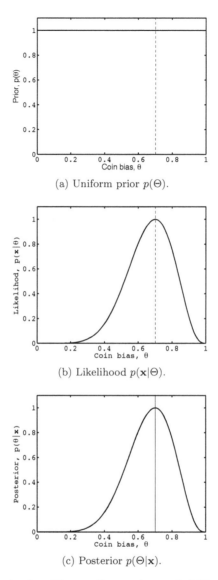

(a) Uniform prior $p(\Theta)$.

(b) Likelihood $p(\mathbf{x}|\Theta)$.

(c) Posterior $p(\Theta|\mathbf{x})$.

Figure 4.4: Bayes' rule with a uniform prior probability density function. **a)** The uniform prior pdf $p(\Theta)$ for coin bias Θ. **b)** The likelihood function $p(\mathbf{x}|\Theta)$ for $x = 7$ heads (out of 10 flips). **c)** Multiplication of corresponding heights on the prior pdf $p(\Theta)$ (a) and the likelihood function $p(\mathbf{x}|\Theta)$ (b) is a scaled version of the posterior pdf $p(\Theta|\mathbf{x})$ (c). The uniform prior pdf guarantees that the likelihood function and the posterior pdf have exactly the same shape, so $\theta_{MLE} = \theta_{MAP}$. For display purposes, graphs (b) and (c) have been scaled to a maximum value of one.

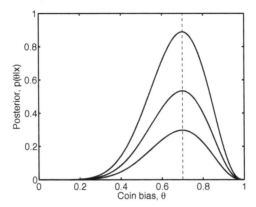

Figure 4.5: The location of the maximum is not affected by constants. The posterior distribution in Figure 4.4c is re-plotted here using 3 different scaling factors applied to Equation 4.19, which have no effect on the location (vertical line) of the maximum, the MAP estimate of the true value of θ.

in the likelihood function is the same as the location of the peak in the posterior probability density function. As these peaks correspond to the maximum likelihood estimate (MLE) and the maximum a posteriori (MAP) estimate of the true value of Θ, this means that $\theta_{MLE} = \theta_{MAP}$.

Putting all this together, the posterior probability that $\Theta = \theta$ is

$$p(\theta|\mathbf{x}) \quad = \quad p(\mathbf{x}|\theta)p(\theta)/p(\mathbf{x}) \tag{4.18}$$

$$= \quad \theta^x(1-\theta)^{N-x}, \tag{4.19}$$

where we have disregarded the constant $p(\mathbf{x})$ in Equation 4.19.

As was the case with the likelihood function (p82), the posterior probability distribution $p(\Theta|x)$ can be plotted by using Equation 4.19 to evaluate $p(\theta|x)$ over a range of 'try out' values of θ, as in Figure 4.4c. The value (0.7) of θ that is most probable, given that $7/10$ heads were observed, corresponds to the peak of Figure 4.4c, and is the MAP estimate of θ_{true}.

MAP Estimates Are Not Affected By Constants

The main reason for obtaining an equation for the posterior distribution $p(\Theta|x)$ is to find the MAP estimate θ_{MAP} of the (unknown) true value θ_{true}, where θ_{MAP} is the value of θ that corresponds to the maximum (peak) in the posterior distribution $p(\Theta|x)$. As an example, the posterior distribution shown in Figure 4.4c has been re-plotted in Figure 4.5 using 3 different values of an arbitrary scaling factor c. The main thing to notice is that the location of the maximum does not change for different values of c, even though the height of the maximum varies. Thus, re-scaling the posterior distribution has no effect on the location of the maximum, which remains at $\theta = 0.7$.

This is a general lesson for finding the maximum (or minimum) value of a parameter like θ: any terms that do not depend on the value of θ (eg $p(\mathbf{x})$) can be treated as constants, and essentially ignored for the purposes of finding the maximum value. Moreover, this lesson applies equally to posterior distributions and to likelihood functions, which is why we can discard constants when seeking the MLE. It also justifies re-scaling many graphs in this book so that they have a maximum value of one. Finally, a similar line of reasoning implies that additive constants can also be ignored.

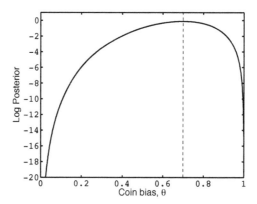

Figure 4.6: The posterior distribution in Figure 4.5 (top curve) is re-plotted here as a log posterior distribution. This changes the shape of the distribution, but has no effect on the location (vertical line) of the maximum.

4.6. Finding the MAP Analytically

We can find the MAP by exhaustive search over all possible values Θ of θ, as above. Alternatively, if we have an equation for the posterior probability distribution then we can find the MAP analytically.

This is much easier if we first find the logarithm of the posterior distribution, as explained below. Taking the logarithm of any function (eg posterior probability distribution) has no effect on the location of its maximum (or minimum), even though (unlike the simple re-scaling above) it alters the overall shape of the function, as shown in Figure 4.6. In order to proceed we need to note that, given two numbers a and b, the logarithm of $(a \times b)$ is

$$\log(a \times b) = (\log a) + (\log b), \tag{4.20}$$

and if each of these numbers is raised to some power $a^{\alpha}b^{\beta}$ then

$$\log(a^{\alpha}b^{\beta}) = (\alpha \log a) + (\beta \log b). \tag{4.21}$$

Similarly, taking the logarithm of Equation 4.19 yields the *log posterior probability*

$$\log p(\theta|x) = x \log \theta + (N - x) \log(1 - \theta). \tag{4.22}$$

The derivative of the log posterior is

$$\frac{d \log p(\theta|x)}{d\theta} = \frac{x}{\theta} - \frac{N - x}{1 - \theta}. \tag{4.23}$$

Given that the derivative is the slope of the distribution shown in Figure 4.6, and that this slope is zero at its maximum value, it follows that the derivative is equal to zero at θ_{MAP}. So if we set the derivative to zero then we can replace θ with θ_{MAP}

$$\frac{d \log p(\theta|x)}{d\theta} = \frac{x}{\theta_{MAP}} - \frac{N - x}{1 - \theta_{MAP}} \tag{4.24}$$

$$= 0, \tag{4.25}$$

and then re-arrange this to solve for θ_{MAP}, which yields $\theta_{MAP} = x/N$. So, for a uniform prior, the MAP estimate θ_{MAP} of the true value θ_{true} is given by the proportion of heads observed in N coin flips.

Using calculus to find the derivative of the sum of terms (as in Equation 4.22), is much easier than finding the derivative of the product of terms (as in Equation 4.19); this is the main reason for working with the log posterior probability rather than posterior probability (and the same reasoning applies to likelihoods and log likelihoods). As explained above, both functions have a maximum θ_{MAP} at the same value of θ, so we can use the log posterior to estimate θ_{MAP}.

4.7. Evolution of the Posterior

Here, we consider how the posterior distribution of coin bias values changes for successive coin flip outcomes (with a uniform prior distribution), as shown in Figure 4.7.

If the first coin flip yields a head, represented as x_1, then, using the same line of reasoning as in Section 4.1, the posterior distribution is

$$p(\Theta|x_1) \quad = \quad \frac{p(x_1|\Theta)p(\Theta)}{p(x_1)} \qquad (4.26)$$

$$= \quad \Theta^1(1-\Theta)^{1-1} \qquad (4.27)$$

$$= \quad \Theta, \qquad (4.28)$$

as in Figure 4.7a. If the second coin flip yields a tail x_2 then

$$p(\Theta|x_1, x_2) \quad = \quad \frac{p(x_1, x_2|\Theta)p(\Theta)}{p(x_1, x_2)} \qquad (4.29)$$

$$= \quad \Theta(1-\Theta). \qquad (4.30)$$

Figure 4.7: (Opposite) The posterior binomial probability distribution $p(\Theta|x)$ of different coin bias values Θ for different numbers N of flips of a coin with bias $\theta_{true} = 0.6$ (dashed line).
a) Probability distribution $p(\Theta|x_1)$ for a coin that lands heads up.
b) Probability distribution $p(\Theta|x_1, x_2)$ for two coin flips (a head and a tail).
c) Probability distribution $p(\Theta|x_1, x_2, x_3, x_4)$ for four coin flips, where three land heads up. Graphs have been scaled to a maximum value of 1.

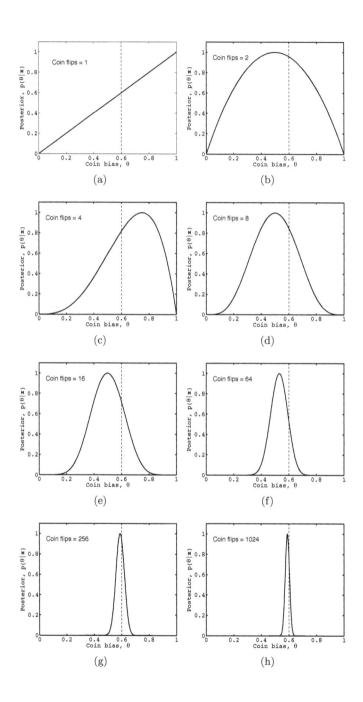

Code Example 4.4: Posterior Evolution
File: Ch4Fig7.py

```python
# Make vector of possible coin bias values, bias = p(head).
tiny = 1e-12
b = np.linspace(0, 1, 1000) # Vector of bias values.
a = 1-b                      # Vector of p(tails) values.


# Choose prior, either uniform or binomial.
useuniformprior = 1
if useuniformprior:
    prior=b**0          # Make prior uniform.
else:
    prior=b**2 * a**2 # Make prior binomial.


# Make data x0, from which individual data set x will be extracted.
coinbias=0.6
ni = 10
NN = 2**ni # Max number of coin flipoutcomes in loop below is NN.
flipoutcomes = np.random.rand(NN);      # Get NN random numbers.
x0 = np.random.rand(NN,1) < coinbias    # flip coin NN times.
x0[0]=1; x0[1]=0; x0[2]=1; x0[3]=1      # cheat initial outcomes!
x0=x0.astype(int)                       # Convert bool to integers.


# Find likelihood and posterior for different numbers N of flips.
# Make container figure for sub-plots.
fig1 = plt.figure(1,(10,16))
for i in range(0, ni):  # Increase coin flips in powers of 2.
    N = 2**i            # Get number of coin flips.
    x = x0[0:N]         # Extract data from x0 for first N flips.
    k = sum(x)          # Get number of heads.
    nh = k              # nh = number of heads.
    nt = N-k            # nt = number of tails.
    #   Likelihood function = probability of nh heads and nt tails.
    lik = b**nh * a**nt
    p = lik*prior       # Find posterior distribution.
    maxp = max(p)       # Find max value of p.
    p = p/maxp          # Make max value of p be one.
    ind = p.argmax()    # Find index of max value of p.
    best = b[ind]       # This is the MAP estimate of bias.
    #   Plot posterior distribution.
    fig2 = plt.subplot(ni/2,2,i+1)
    plt.plot(b,p)
    plt.xlabel("Coin Bias, $\\theta$")
    plt.ylabel("Posterior $p(\\theta|x)$")
    plt.ylim((0,1.1))
    plt.text(0.05, 0.9, 'Num flipoutcomes = ' + str(N))
    plt.text(0.05, 0.8, 'Num heads = ' + str(nh))
    plt.text(0.05, 0.7, 'Bias est = '+'%(num)1.3f'% {"num" : best})
plt.show()
```

Again, the value of $p(\theta|x_1, x_2)$ has been plotted in Figure 4.7b for a range of putative values of coin bias values θ between zero and one. This process has been repeated for increasing numbers of coin flips, and the peak of the posterior distribution moves closer to the true value of $\theta_{true} = 0.6$ as N increases, as shown in Figure 4.7.

Error Bars for Binomials

As the number N of coin flips increases, the width of the posterior distribution shrinks, and so our confidence in the value of Θ at its peak increases. However, we would like to know precisely how much the width shrinks as N increases, where width is defined as the *standard deviation* (see Appendix A for a definition of these terms). The standard deviation of the posterior distribution in Equation 4.19 is[38]

$$\sigma_N = \sqrt{\frac{\theta_{MAP}(1 - \theta_{MAP})}{N}}. \qquad (4.31)$$

This says that, as the number N of coin flips increases, the standard deviation σ_N shrinks in proportion to $1/($the square root of $N)$, and, as σ_N decreases so our confidence in the value of θ_{MAP} increases. This can be seen qualitatively in Figure 4.7 in the shrinking width of the posterior distribution. Equation 4.31 is a useful result, which applies in a wide variety of contexts beyond flipping coins.

Sequential Inference

We have already seen how the combined set of outcomes can be used to infer coin bias. However, just as new evidence may arrive in a court of law, so each coin flip provides additional evidence regarding the coin bias. Rather than waiting until all outcomes have been observed, we can use the posterior distribution from all outcomes up until the last coin flip as a prior distribution for interpreting the next flip outcome.

For two outcomes x_1 and x_2, the posterior probability distribution is

$$p(\Theta|x_1, x_2) = \frac{p(x_1, x_2|\Theta)p(\Theta)}{p(x_1, x_2)}. \qquad (4.32)$$

If successive outcomes are mutually independent, so that the outcome of the second flip does not depend on the outcome of the first flip then the joint probability *factorises* so that

$$p(x_1, x_2) \quad = \quad p(x_1)p(x_2). \tag{4.33}$$

This, in turn, means that the likelihood function factorises, so that

$$p(x_1, x_2|\Theta) \quad = \quad p(x_1|\Theta)p(x_2|\Theta). \tag{4.34}$$

Substituting Equations 4.33 and 4.34 into 4.32 yields

$$p(\Theta|x_1, x_2) \quad = \quad \frac{p(x_2|\Theta)}{p(x_2)} \times \frac{p(x_1|\Theta)p(\Theta)}{p(x_1)}, \tag{4.35}$$

where (according to Equation 4.26) the final ratio in Equation 4.35 is $p(\Theta|x_1)$, so that

$$p(\Theta|x_1, x_2) \quad = \quad \frac{p(x_2|\Theta)p(\Theta|x_1)}{p(x_2)}. \tag{4.36}$$

Thus, the posterior probability distribution $p(\Theta|x_1)$ obtained from the first coin flip can act as a prior for inference regarding the value of Θ after the second coin flip. More generally, if we define a combination of N outcomes as $\mathbf{x}_N = \{x_1, x_2, \ldots, x_N\}$, then there will be between 0 and N heads. As usual, the posterior probability distribution is

$$p(\Theta|\mathbf{x}_N) \quad = \quad \frac{p(\mathbf{x}_N|\Theta)p(\Theta)}{p(\mathbf{x}_N)}. \tag{4.37}$$

If we flip the coin one more time then we have $N+1$ coin flip outcomes $\mathbf{x}_{N+1} = (\mathbf{x}_N, x_{N+1})$, with a posterior probability distribution

$$p(\Theta|\mathbf{x}_{N+1}) \quad = \quad \frac{p(\mathbf{x}_{N+1}|\Theta)p(\Theta)}{p(\mathbf{x}_{N+1})}. \tag{4.38}$$

Now, using the same logic as in Equation 4.36, we can use the posterior $p(\Theta|\mathbf{x}_N)$ as a prior probability for the next coin flip outcome, x_{N+1}, and we can then use this to update the posterior

$$p(\Theta|\mathbf{x}_{N+1}) \quad = \quad \frac{p(x_{N+1}|\Theta)p(\Theta|\mathbf{x}_N)}{p(x_{N+1})}, \qquad (4.39)$$

and this can be repeated for each successive outcome. Note that the mutual independence of coin flip outcomes guarantees that the posterior distribution is identical whether it is computed using all of the data simultaneously (Equation 4.38) or sequentially (Equation 4.39).

4.8. Reference Priors

The question of what constitutes a fair prior for an unknown parameter lies at the heart of Bayesian analysis. Both Bayes and Pascal assumed that we should assign an equal (ie uniform) prior probability to all parameter values. In essence, a uniform prior was deemed fair because it expresses our state of ignorance regarding a parameter. Indeed, it could be argued that maximum likelihood estimation is justified because it is equivalent to Bayesian inference with a uniform prior.

However, it can be shown that a uniform prior, used explicitly (as part of Bayes' rule) or implicitly (ie via maximum likelihood estimation), does not necessarily represent the most 'fair' or unbiased assumption regarding the prior distribution. This was eloquently summarised by Jeffreys [19] on the first page of this chapter.

The idea of expressing a particular degree of ignorance fairly in the form of a prior distribution has a long history. It began life as *Bayes' postulate*, and was expressed by Laplace(1749-1827) in a form that came to be known as the *principle of insufficient reason*, which was renamed by the economist John Keynes(1921) as the *principle of indifference*. This general approach, which seeks to preclude any form of bias, was championed by Jeffreys(1939) [19], and is known as *objective Bayes*. These fair priors are also known as *uninformative* or *non-informative* priors. Bernardo(1979) [3] coined the term *reference prior*, and Jaynes(2003) [18] expressed the idea more generally as the *principle of maximum entropy*, or *maxent*.

Maxent is based on the simple idea that the prior distribution should be consistent with any information we do have, but in all other respects, it should provide as little information as possible regarding the posterior distribution of the parameter. For example, if we know that the data is a sample from an exponential distribution then it can be shown that the reference prior distribution for the mean θ is $p(\theta) \propto 1/\theta$. By implication, any prior except for the reference prior would bias the posterior distribution of parameter values in accordance with prior information that we do not have. A brief account of these ideas, which are a topic of active research, is given in Appendix H.

Bootstrapping

When it is not obvious what form a fair prior distribution should take in order to obtain the posterior distribution, a computer-intensive method called *bootstrapping* can be used. Briefly, this uses repeated sampling of the data to construct an estimate of the posterior distribution. Crucially, this estimated posterior is similar to the one that would have been obtained if we had used a reference (ie fair) prior[11].

4.9. Loss Functions

As we have seen, Bayesian inference provides a posterior distribution $p(\Theta|\mathbf{x})$, based on data $\mathbf{x}$. As we are often interested in estimating a value for Θ, we can use this posterior to choose a single value or *point estimate* $\hat{\theta}$ (pronounced *theta hat*). However, the particular point estimate we choose depends on how much we care about inaccuracies in $\hat{\theta}$. So far, we have used $\hat{\theta} = \theta_{MAP}$ as the 'natural' choice for a point estimate, but we have not justified this choice.

The value of $\hat{\theta}$ is always based on noisy data $\mathbf{x}$, so $\hat{\theta}$ contains some error, and if the true value is θ then this error is $\Delta = (\hat{\theta} - \theta)$. If the cost of choosing the wrong value of θ rises sharply as the error increases then this should somehow be reflected in our choice of $\hat{\theta}$. For example, if Θ represents a range of possible diseases then the cost of getting its value wrong rises very steeply, and we could end up treating a patient who has smallpox as if he merely had chickenpox. Such costs are defined by a *loss function*, where minimising losses amounts to maximising gains.

We can explore the consequences of using different loss functions by considering the different costs used in an archery contest. However, for technical reasons (to do with medians), we restrict the target to lie on a vertical line. We will assume your aim is so good that you can hit anything you aim for, so if you aim for a point $\hat{\theta}$ then that is exactly where your arrow will land. But, as your aim is based on noisy visual data $\mathbf{x}$, your knowledge of the bullseye's probable location is summarised by the posterior pdf $p(\Theta|\mathbf{x})$, which defines the probability that the bullseye is in each possible location θ. The value of $\hat{\theta}$ that minimises the expected posterior loss is called a *Bayes estimate*.

Zero-one Loss Function: Suppose the contest rules state that if you hit the bullseye then you lose zero points, but if you miss then you lose one point (so your objective is to lose as few points as possible). Using this *zero-one loss function*, it can be shown[27;30] that the way to minimise your average or *expected posterior loss* (ie averaged over values of θ defined by the posterior distribution $p(\Theta|\mathbf{x})$) is to set $\hat{\theta}$ to be the *mode* (ie maximum) of $p(\Theta|\mathbf{x})$, so that $\hat{\theta} = \theta_{MAP}$. In other words, you should aim for the point which corresponds to the most probable value of θ given $\mathbf{x}$.

Quadratic Loss Function: If the rules state that the number of points you lose is equal to the square of the arrow-bullseye distance then this defines a *quadratic loss function*. In this case, if the arrow lands at the location $\hat{\theta}$ and the bullseye is at θ then the loss incurred is

$$\Delta^2 = (\hat{\theta} - \theta)^2. \tag{4.40}$$

If an average is taken over all possible continuous values of arrow-bullseye distances then the expected posterior loss is

$$\mathrm{E}[\Delta^2] = \int_\theta p(\theta|\mathbf{x})\,(\theta - \hat{\theta})^2\,d\theta, \tag{4.41}$$

where this average implicitly provides a weighting defined by the posterior probability that the bullseye is in each location. It can be shown [27;30] that the way to minimise the average loss for a quadratic loss function is to set the value of the Bayes estimate $\hat{\theta}$ to be the mean of $p(\Theta|\mathbf{x})$ (see Appendix D, p160). Thus, if you wish to minimise the average squared error then you should aim for the point which corresponds to the mean of the posterior pdf.

Absolute Loss Function: If the rules state that the number of points you lose is the arrow-bullseye distance then this defines an *absolute loss function*. In this case, it can be shown [27] that the way to minimise the expected posterior loss is to set the value of the Bayes estimate $\hat{\theta}$ to be the median of $p(\Theta|\mathbf{x})$, so you should aim for the point which corresponds to the median of the posterior (for an ordered set of numbers, the median is the number in the middle).

In practice, it may make little difference which loss function is used because the posterior distribution is often Gaussian (see Chapter 5), which means that it is symmetric for a given parameter. This symmetry guarantees that the mean, median and mode have the same value, so all three loss functions provide the same estimate.

We will not explore loss functions further, but it is worth noting that, when it is claimed that Bayesian inference is optimal, this optimality is always defined with respect to a particular loss function.

Summary

We have explored how to use Bayes' rule to estimate the value of a continuous parameter, the bias associated with a coin being flipped. We derived an equation for the posterior pdf, so that our estimate of the true parameter value could be based on analytic methods. Demonstrations of Bayesian inference for coin bias using a uniform and non-uniform prior were given. For the first time, we confronted the issue of how to choose a prior for inference, and we considered a problem for which a uniform prior is the wrong prior to use.

Chapter 5

Gaussian Parameter Estimation

> The true logic for this world is the calculus of Probabilities,
> which takes account of the magnitude of the probability
> which is, or ought to be, in a reasonable man's mind.
> James Clerk Maxwell, 1850.

Introduction

This chapter is intended as an introduction to parameter estimation and
regression. The concepts introduced here are by no means exclusively
Bayesian, but they are fundamental to more advanced statistical
analysis, Bayesian or otherwise. We begin with an exploration of a
ubiquitous distribution, the *Gaussian* distribution.

5.1. The Gaussian Distribution

If the heights of many people are plotted as a histogram then we obtain
the typical bell shape shown in Figure 5.1a, which has approximately
the same shape as the Gaussian distribution shown in Figure 5.1d. This
distribution is defined by two parameters, its *mean* (centre) and width
or *standard deviation* (see Appendix F).

If the population comprises $N = 5,000$ people, and the ith person
has a height x_i, then using the summation convention (Appendix B),

the mean height μ_{pop} (mu) in the population is

$$\mu_{pop} \;=\; \frac{1}{N}\sum_{i=1}^{N} x_i, \tag{5.1}$$

and the standard deviation of these heights is

$$\sigma_{pop} \;=\; \left(\frac{1}{N}\sum_{i=1}^{N}(x_i - \mu_{pop})^2\right)^{1/2}. \tag{5.2}$$

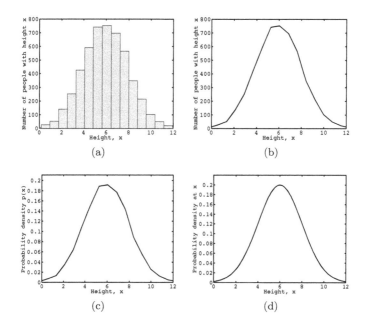

(a)　　　　　　(b)

(c)　　　　　　(d)

Figure 5.1: Histograms and probability density functions.
a) A histogram of N values of height x, measured at each of N_x heights $(x_1, x_2, \ldots, x_{N_x})$. This histogram was constructed by dividing values of x into N_x bins, and plotting the number of height values in each bin. **b**) Histogram outline, found by joining up the heights of adjacent bins. **c**) Outline obtained after (b) was normalised to have an area of one. This approximates the probability density function (pdf) $p(x)$ of x. **d**) A Gaussian pdf provides a good approximation to (c).

We should note that the variance is defined as the square of the standard deviation σ_{pop}^2. This equation looks complicated, but close inspection reveals that it is actually the square root of [the average squared difference between the population mean μ_{pop} and each data point x_i], so it's a measure of the variability in x.

For convenience, we will assume that the distribution of heights in the population is Gaussian, with (unknown) mean μ_{pop} and (known) variance σ_{pop}^2. Now, this Gaussian assumption means that the probability that a randomly chosen person has a height x_i is defined by the equation for a Gaussian distribution

$$p(x_i|\mu_{pop}, \sigma_{pop}) \quad = \quad k_{pop} \, \exp \frac{-(x_i - \mu_{pop})^2}{2\sigma_{pop}^2}. \qquad (5.3)$$

The exponential function exp for any quantity, such as z, is defined as $\exp(z) = e^z$, where e is the constant 2.718. (Strictly, we should be speaking of the probability *density* of x_i, see Appendix D). The constant k_{pop} ensures that the Gaussian distribution has an area of one $k_{pop} = 1/(\sigma_{pop}\sqrt{2\pi})$, and therefore the probabilities it defines sum to one. Note that both x and μ_{pop} are continuous parameters.

5.2. Estimating the Population Mean

Suppose we wish to find the average height μ_{pop} of a population of N people, based on a random sample of n people, where $n < N$. The data in this sample of n heights is

$$\mathbf{x}_n \quad = \quad (x_1, x_2, \ldots, x_n), \qquad (5.4)$$

and the mean of our sample of n heights is

$$\mu_{est} \quad = \quad (1/n) \sum_{i=1}^{n} x_i, \qquad (5.5)$$

where the subscript *est* stands for *estimate*, because this is an estimate of the population mean. In this example, we know the population variance σ^2_{pop}, but if we did not, then we could use the sample variance as an estimate of it. The sample variance is

$$\sigma^2_{est} \quad = \quad \frac{1}{n} \sum_{i=1}^{n} (x_i - \mu_{est})^2. \tag{5.6}$$

The Likelihood Function

The likelihood function tells us the probability of the data given some putative value μ of the population mean μ_{pop}. Each height in our sample was chosen at random from the population of heights, so the probability that we chose the height x_i is given by the likelihood function for x_i in Equation 5.3, repeated here with μ_{pop} replaced by μ

$$p(x_i|\mu) \quad = \quad k_{pop} \, e^{-(x_i-\mu)^2/(2\sigma^2_{pop})}. \tag{5.7}$$

As the sample of n heights is chosen from a population of N heights, the mean of this sample is a random variable, which has a Gaussian distribution with mean μ_{pop}, and a standard deviation[1] of $\sigma_n = \sigma_{pop}/\sqrt{n}$ (Section 5.3, and Appendix A). Since the sampled heights are mutually independent, we can multiply their individual probabilities (Equation 5.7) to obtain the probability of the data given a putative population mean μ, which is called the likelihood of μ

$$p(\mathbf{x}|\mu) \quad = \quad \prod_{i=1}^{n} p(x_i|\mu) \tag{5.8}$$

$$= \quad \prod_{i=1}^{n} k_{pop} \, e^{-(x_i-\mu)^2/(2\sigma^2_{pop})}, \tag{5.9}$$

where the capital Greek letter Π (pi) is an instruction to multiply all the terms to its right with each other (see Appendix B).

When considered over a range of plausible values of μ, Equation 5.9 defines a *likelihood function* for a given sample of n heights. In the process of estimating the population mean μ_{pop}, we try out different

---

[1] The standard deviation of a set of means is the *standard error*, by convention.

putative values μ for the population mean. Each value defines a different model for the observed data, and some models fit the data better than others, where this fit is measured in terms of how probable each model makes the observed data. The model parameter value (μ) that makes the observed data most probable is, by definition, the MLE.

The Posterior Probability Density Function

If all values of the mean μ are equally probable (within a specific range) then this defines a uniform prior distribution $p(\mu)$, for which the posterior probability density of a value μ is given by Bayes' rule

$$p(\mu|x_i) \quad = \quad p(x_i|\mu)p(\mu)/p(x_i) \qquad (5.10)$$

$$= \quad c_\mu\, k_{pop}\, e^{-(x_i-\mu)^2/(2\sigma^2_{pop})}, \qquad (5.11)$$

where we can treat $c_\mu = p(\mu)/p(x_i)$ as constant for a given observation x_i. Since the measured heights are mutually independent, we can multiply their individual posterior probability densities together to obtain the posterior probability density of a putative value μ for the population mean, based on n values of x_i

$$p(\mu|\mathbf{x}) \quad = \quad \prod_{i=1}^{n} c_\mu\, k_{pop}\, e^{-(x_i-\mu)^2/(2\sigma^2_{pop})}. \qquad (5.12)$$

When considered over a range of different values of μ, Equation 5.12 defines the posterior probability density function (pdf), which has been plotted for different sample sizes (values of n) in Figure 5.2. Notice how the width of the distribution shrinks as n increases, which is a topic we revisit below.

From Posterior Probabilities to Least-Squares Estimation

It is customary to take the logarithm of quantities like those in Equation 5.9 and 5.12 (see Section 4.6), and we will use the latter as an example. As a reminder, given two numbers a and b, $\log(a \times b) = \log a + \log b$,

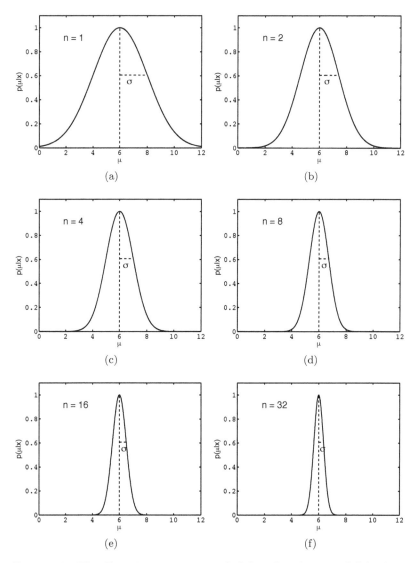

Figure 5.2: The Gaussian posterior probability distribution $p(\mu|\mathbf{x})$ of the mean μ_{pop} for different sample sizes $n = 1, 2, 4, 8, 16$ and 32 for the data $\mathbf{x}$ drawn from a population with mean $\mu_{pop} = 6$ and standard deviation $\sigma_{pop} = 2$. The standard deviation σ_n in the distribution of mean values for a sample size n is proportional to $1/\sqrt{n}$. Each horizontal line has length σ_n. Graphs have been re-scaled to a maximum value of one.

so that the log posterior distribution is defined by

$$\log p(\mu|\mathbf{x}) \quad = \quad \log \prod_{i=1}^{n} p(\mu|x_i), \tag{5.13}$$

$$= \quad \sum_{i=1}^{n} \log c_\mu \, k_{pop} \, e^{-(x_i-\mu)^2/(2\sigma_{pop}^2)}, \tag{5.14}$$

and if we sum over individual terms then we have

$$\log p(\mu|\mathbf{x}) \quad = \quad \sum_{i=1}^{n} \log c_\mu \, k_{pop} + \sum_{i=1}^{n} \log \, e^{-(x_i-\mu)^2/(2\sigma_{pop}^2)} \tag{5.15}$$

$$= \quad \kappa - \frac{1}{2} \sum_{i=1}^{n} \frac{(x_i - \mu)^2}{\sigma_{pop}^2}, \tag{5.16}$$

where the Greek letter κ (kappa) represents the summed constants on the left side of Equation 5.15. At this point, you may recognise this as being similar to a *least-squares* operation, which it is, almost. The value of the additive constant κ is irrelevant, because it makes no difference to our final estimate (see Section 4.5), so we can simply ignore it. We should acknowledge this cavalier action by defining a new function

$$G \quad = \quad -\frac{1}{2} \sum_{i=1}^{n} \frac{(x_i - \mu)^2}{\sigma_{pop}^2}. \tag{5.17}$$

Note that, because all terms are squared, they are all positive, so that G must be negative. The value of μ that makes Equation 5.17 as large as possible (but always negative) also makes its negative as small as possible (but always positive). Thus, the value of μ that *maximises* Equation 5.16 also *minimises* its negative, for which we define a related function $F = -G$

$$F \quad = \quad \frac{1}{2} \sum_{i=1}^{n} \frac{(x_i - \mu)^2}{\sigma_{pop}^2}. \tag{5.18}$$

This looks more like a least-squares operation. We know that σ_{pop} is the same for each measurement x_i so its value has no effect on the final estimate of the mean (see Section 4.5, p91). We can therefore multiply each term in the sum of Equation 5.18 by $2\sigma_{pop}^2$ (even though we do not know its value), which yields $(E = F2\sigma_{pop}^2)$

$$E \quad = \quad \sum_{i=1}^{n}(x_i - \mu)^2. \tag{5.19}$$

This, we can recognise as the 'standard' least-squares operation used in conventional statistics. Specifically, the value of μ that minimises Equation 5.19 is the *least-squares estimate* (LSE) of the population mean μ_{pop}. As this was derived using Bayesian methods, it implies that, given certain assumptions, the Bayesian estimate of the population mean is identical to the least-squares estimate. These assumptions are that, 1) the prior distribution is uniform, 2) the measurements are mutually independent, and 3) the standard deviation of the noise is the same for all measurements.

Given these assumptions, the posterior probability density function of Equation 5.12 reduces to the least-squares expression in Equation 5.19. It can be shown (Appendix G) that the estimated population mean implied by the least-squares estimate (Equation 5.19), and therefore by the posterior distribution in Equation 5.12, are both

$$\mu_{est} \quad = \quad (1/n)\sum_{i=1}^{n}x_i, \tag{5.20}$$

which is just the mean of our sample. So it turns out that, given certain assumptions, the least-squares estimate is equal to the MAP estimate of the population mean, and that both are given by the sample mean.

However, before we did this analysis, we could not have known that the result would come out as it did. By deriving this result using a rigorous Bayesian analysis, we can be confident that the sample mean is not an *ad hoc* estimate of the population mean.

5.3. Error Bars for Gaussian Distributions

It seems plausible that our estimate of the population mean should become more accurate as the sample size n increases (as we noted in

Chapter 4). But how should this increased accuracy be measured? Briefly, our confidence in the estimated mean increases as the standard deviation σ_n of the posterior distribution decreases, such that

$$\sigma_n \quad = \quad \sigma_{pop}/\sqrt{n}. \tag{5.21}$$

This states that the standard deviation of sample means is proportional to $1/$(the square root of the number n of data points), as shown in Figure 5.2. Our confidence in the estimated mean is expressed as

$$\mu_{true} \quad = \quad \mu_{est} \pm \sigma_{pop}/\sqrt{n}, \tag{5.22}$$

which states that the true value of the population mean is estimated to be μ_{est} with *error bars* defined by $\sigma_{pop}/\sqrt{n}$.

Error Bars and the Distribution of Means

Supposing we estimate the mean many times using a sample size n. A well-known theorem called the *central limit theorem* (see below) ensures that, if we plot a histogram of these estimated means then we obtain a distribution that is approximately Gaussian. The standard deviation of this distribution of means is called the *standard error in the mean* (sem). Crucially, this distribution of sample means is an approximation to the posterior distribution of the population mean in Equation 5.11.

As we increase the number n of data points in each sample, the sample means become increasingly similar to the population mean, and therefore to each other, and so the standard deviation of the distribution of sample means decreases. This can be seen qualitatively in Figure 5.2 from the shrinking width of the posterior distribution as n increases. According to Equation 5.21, the standard deviation of these distributions shrinks in proportion to $1/$(the square root of n).

The Central Limit Theorem

The result presented in Equation 5.21 does not apply only to data with a Gaussian distribution of values, but has an extraordinarily wide range of application. This is because, in essence, the central limit theorem states that, as the number n of points in each sample from almost any distribution increases, the distribution of mean values becomes increasingly Gaussian.

Because almost every physical quantity (eg human height) is affected by multiple factors, it is effectively a weighted mean, which suggests that such quantities should have a Gaussian distribution. This may account for the ubiquity of the Gaussian distribution in nature, and provides a rational basis for the default assumption that distributions are Gaussian.

5.4. Regression as Parameter Estimation

Suppose we have a vague suspicion that tall people have higher salaries than short people. How can we test this hypothesis? Well, we could obtain the salaries of, say $n = 11$ people, and plot a graph of income versus height, as in Figure 5.3. If the line we draw has a positive slope then this suggests that salary and height are related. But how do we know where to draw the line? The solution is given by a procedure called *linear regression*, which involves fitting a straight line to data.

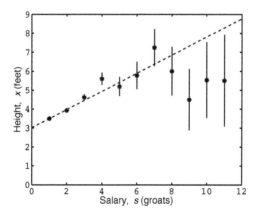

Figure 5.3: Regression. The heights x_i of 11 people plotted against their salaries s_i, and fitted with a line. The line's slope is $m = 0.479$ and the intercept is $c = 3.02$, so that regressed height is $\hat{x}_i = 0.479 \times s_i + 3.02$. Each vertical line's length is the known standard deviation of the corresponding data point, which increases with height in this example. These data points were generated from noisy measurements of a line with slope $m = 0.5$ and intercept $c = 3$.

Code Example 5.1: Regression as Parameter Estimation
File: Ch5Eq26.py (Figure 5.3)

```
# Make noisy data with slope m and intercept c.
# Define 11 salary values for horizontal axis.
s = arange(1,12)
# Set value of slope m and intercept c.
m = 0.5; c = 3
# Set standard deviation of each measured height.
sds = 2*arange(1,12)/10.0; sds = sds * arange(1,12)/10.0
# Use noise values copied from book (based on sds above).
eta=[-0.0023,-0.0728,0.1104,0.6076,-0.3034,-0.2237,0.7407,
-1.0,-3.0,-2.4653,-3.0]
# Find observed values of x (with noise added).
x = m*s + c + eta

# Weighted Least Squares regression.
# Find weightings w (discount) for each data point.
vars0 = sds ** 2; w = 1/vars0
# Un-comment next line for solution based on un-weighted regression.
#w=ones(size(w))
ss = sm.add_constant(s) # Add column of 1s for regression.
model = sm.WLS(x,ss, weights=w)
results = model.fit()
cest2, mest2 = results.params
print('Estimated␣slope␣=␣%.3f,' % mest2)
print('␣estimated␣intercept␣=␣%.3f.' % cest2)

# Make line xest2 based on fitted slope and intercept.
s2 = arange(0,13)
xest2 = mest2 * s2 + cest2

# Plot fitted line xest, data points, and error bars.
fig1 = plt.figure()
plt.errorbar(s, x, yerr=sds, fmt='o',color='k')
plt.plot(s,x, 'k*', s2, xest2, 'k--')
plt.xlabel('Salary,␣$s$␣(groats)')
plt.ylabel('Height,␣$x$␣(feet)')
plt.xlim((0,12)); plt.ylim((0,9))
# Output:
#    Estimated slope = 0.479.
#    Estimated intercept = 3.019.
```

Note that, in order to use regression, we assume that salary can be measured exactly, but that height measurements contain some jitter or noise. This noise may be due to an inability to measure height exactly, for example. We will refer to the jitter in the data as *measurement noise*. The uncertainty this noise generates means that we treat

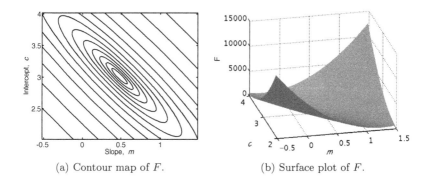

(a) Contour map of F.   (b) Surface plot of F.

Figure 5.4: a) Contour map of the values of F, for different putative values of slope m and intercept c. b) Surface plot of F. Both graphs are centred on the values $m = 0.479, c = 3.02$ associated with the minimal value of F.

height as a random variable, and we assume that the noise in height measurements has a Gaussian distribution. In this example, we assume that the noise in measurements increases with height.

In the case of regression, the parameters we seek to estimate specify the characteristics of the line we fit to our data, as shown in Figure 5.3. For a straight line, the relevant parameters are its *slope* m and *intercept* c, which specify how the estimated height $\hat{x}$ (x *hat*) increases with salary s, $\hat{x}_i = ms_i + c$, where, by definition, each value $\hat{x}_i$ lies on the fitted line.

This equation represents a *model* (with parameters m and c) of how height x increases with salary s. The vertical distance between the measured height and the idealised height $\hat{x}_i$ that conforms to the 'straight line' model, as in Figure 5.3, is represented by the Greek letter η (*eta*), such that $\eta_i = x_i - \hat{x}_i$, where η_i is the amount of noise in the value x_i. For simplicity, we assume a uniform joint prior $p(m, c)$. If η has a Gaussian distribution then the posterior distribution of m and c for a measurement x_i is

$$p(m, c | x_i) \quad = \quad k_i \, e^{-\eta_i^2/(2\sigma_i^2)} \tag{5.23}$$

$$= \quad k_i \, e^{-[x_i - (ms_i + c)]^2/(2\sigma_i^2)}, \tag{5.24}$$

5.4. Regression as Parameter Estimation

Code Example 5.2: Regression as Parameter Estimation: Contour Maps
File: Ch5Eq26.py (Figure 5.4a+b)

```
# Code for 2d graph, Figure 5.4a.
# Continued from previous snippet ...
m = mest2; mmin = mest2-1; mmax = mest2+1
c = cest2; cmin = cest2-1; cmax = cest2+1
minc = (mmax-mmin)/100
cinc = (cmax-cmin)/100

Fs = []
ms = linspace(mmin, mmax, 100);    nm = len(ms)
cs = linspace(cmin, cmax, 100);    nc = len(cs)
Farray = zeros((nm,nc))
for m1 in arange(0,nm):
    for c1 in arange(0, nc):
        mval=ms[m1]
        cval=cs[c1]
        y1 = mval*s + cval
        F1 = ((x-y1)/sds) ** 2
        Farray[m1,c1] = sum(F1)
        Fs = (Fs, F1)

fig2 = plt.figure()
Z1 = Farray.T
zmin = amin(Z1); zmax = amax(Z1)
zrange = zmax - zmin
v = arange(zmin, zmax, zrange / 10)
# Adjust spacing of contour lines.
v = arange(0, 9 ,0.5)
v = exp(v)
v = v * zrange/max(v);
X, Y = np.meshgrid(ms,cs)
plt.xlabel('Slope,␣$m$')
plt.ylabel('Intercept,␣$c$')
plt.contour(X, Y, Z1, v)
plt.show()

# Code for 2d graph, Figure 5.4b.
fig3 = plt.figure()
ax = fig3.add_subplot(111, projection='3d')
ax.set_xlim(-0.5, 1.5)
ax.set_ylim(2,4)
ax.set_zlim(min(v), max(v))
ax.autoscale(enable=True, axis='z')
ax.set_xlabel('$m$'); ax.set_ylabel('$c$'); ax.set_zlabel('$F$')
surf = ax.plot_surface(X,Y,Z1,rstride=1,cstride=1,
cmap = cm.coolwarm,linewidth=0,antialiased=True)
plt.show()
```

115

where σ_i is the standard deviation in the noise of the ith height, and the constant is $k_i = 1/(a\sigma_i\sqrt{2\pi})$, where a is the height of the uniform prior $p(m, c)$. If the values of η_i for different measurements are mutually independent then, for n corresponding values $\mathbf{x} = (x_1, \ldots, x_n)$ and $\mathbf{s} = (s_1, \ldots, s_n)$ the joint posterior distribution of m and c is

$$p(m, c|\mathbf{x}) \quad = \quad \prod_{i=1}^{n} k_i \, e^{-[x_i - (ms_i + c)]^2/(2\sigma_i^2)}. \tag{5.25}$$

Following a similar line of reasoning as in Section 5.2 (plus some additional algebra), we obtain the least-squares operation as

$$F \quad = \quad \sum_{i=1}^{n} [(x_i - (ms_i + c))/\sigma_i]^2. \tag{5.26}$$

The standard deviation σ_i of each measured height effectively 'discounts' less reliable measurements, so that noisier measurements have less influence on the parameter values (m, c) of the fitted line.

If we plot the value of F for different putative values of m and c then we obtain the graphs in Figure 5.4, which show the same data, but plotted using different representations. The particular values of m and c that minimise F (ie that make F as small as possible) also maximise the posterior probability (Equation 5.25), and therefore correspond to the MAP estimates of the true slope and intercept, respectively. In practice, we would not normally plot the values of F, but would use a numerical method to find the values of m and c that minimise F.

If the amount of noise associated with all measurements is the same (ie if all σ_i are the same) then its value makes no difference to the final result, and so it can be ignored, which allows us to define

$$E \quad = \quad \sum_{i=1}^{n} [x_i - (ms_i + c)]^2. \tag{5.27}$$

In this case, we could use the least-squares method (Appendix G) to find values of m and c that minimise Equation 5.27. However, if we do this then we effectively ignore (usually because we do not know) the underlying standard deviation of each data point, and we obtain the misleading estimates $m = 0.172, c = 4.18$. This makes the point that we should take account of the reliability of data if it is available.

The particular way we have posed this problem seems to suggest that salary causes increased height, which is clearly silly. However, it does emphasise a crucial fact about regression: a fitted line does not imply causation. This would be less obvious if we had regressed salary against height, rather than vice versa.

The process of fitting a line to a set of data points is qualitatively similar to finding the mean μ of a Gaussian distribution, which consists of finding that value of μ which makes the sum of squared distances between μ and all data points x_i as small as possible (Equation 5.19). Similarly, finding the best fitting line, consists of finding that value of the slope m and intercept c which makes the sum of squared distances between points $\hat{x}_i$ on the fitted line and all data points x_i as small as possible (Equation 5.27).

Summary

The Bayesian method has been applied to estimating one parameter, the mean μ of a Gaussian. It was then applied to several 'means' $\hat{x}_i = mx_i + c$ which lie on a straight line (regression) determined by two estimated parameters, slope m and intercept c. The equivalence between MAP and least-squares estimates was proved.

Chapter 6

A Bird's Eye View of Bayes' Rule

> Every problem becomes very childish when once it is explained to you.
> Sherlock Holmes, from The Dancing Men. AC Doyle, 1905.

Introduction

Now that we have some familiarity with the properties of the Gaussian distribution, we can take a global view of Bayesian analysis. To some extent, this chapter is a repeat of Section 3.5 where we considered discrete variables, whereas we consider continuous variables with a particular (ie Gaussian) continuous joint distribution here.

6.1. Joint Gaussian Distributions

We are going to cover a lot of ground in the next paragraphs, but do not worry about the mathematical details at this stage, because the main objective is to provide an accessible example of a joint distribution.

Imagine a collection of coins, each of which has a different propensity or bias for landing heads up. Let us assume that the mean bias of these coins is $\theta = 0.5$, and, more importantly, we assume that the distribution of biases is Gaussian (Appendix F). This just means that, if we choose a coin at random, then the probability that our coin has a bias θ is governed by a Gaussian distribution. If we flip this coin a number N of times then it can be shown that the number n of heads has a binomial distribution (Appendix E). More importantly, if N

is reasonably large then the central limit theorem (see p111) ensures that n has a distribution that is approximately Gaussian (as briefly described on p164). Finally, if n has a Gaussian distribution then it follows that the distribution of the proportion $x = n/N$ of heads is also Gaussian. Given that the distribution of biases is also Gaussian, the collective values of Θ and X therefore define a 2D *Gaussian joint probability density function* $p(X, \Theta)$, Figure 6.1.

As in previous chapters, we will treat Θ as a parameter whose value θ_{true} we wish to estimate, and x as a value of X that depends on the value of Θ. Here, θ is the bias of a coin, which can be estimated by flipping the coin a number of times to obtain an observed proportion x of heads. For example, if we flip the coin 100 times and observe 75 heads then $x = 0.75$, and our initial estimate of the coin's bias is simply $\theta = 0.75$. In general, x is a reasonably good estimate of θ, so that the values of x and θ are somewhat *correlated*. In essence, this means that, as θ increases, so x also increases. However, as x is the proportion of observed heads which is subject to the effects of chance, it contains some jitter or noise, and is therefore an imperfect estimate of the value of θ. In this example, our coin is chosen from a particular

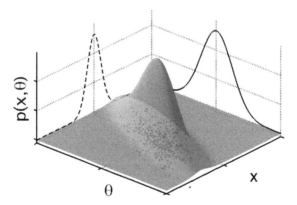

Figure 6.1: Joint probability density function for two correlated variables X and Θ. The local density of dots on the ground plane at any coordinates (x, θ) defines the height $p(x, \theta)$ of the surface above those coordinates. The marginal distributions are the marginal likelihood $p(X)$ (dashed curve) and prior probability density function $p(\Theta)$ (solid curve).

population of coins with different known biases, so the observed value x can be plotted against the known bias θ of our coin. If we repeat this process for every coin in our population of coins then the result is an elongated cluster of points, as shown in the ground plane of Figure 6.1 and in Figure 6.2. This elongated 2D distribution is consistent with the fact that the distribution of coin biases is Gaussian, and that the distribution of the observed proportions of heads is also Gaussian.

Because each pair of values x and θ defines a single location on this ground plane, the local density of dots represents the joint probability density $p(x, \theta)$ of observing any pair $X = x$ and $\Theta = \theta$, which can be represented as the height of a surface above the ground plane, as shown in Figure 6.1 or by intensity as in Figure 6.2.

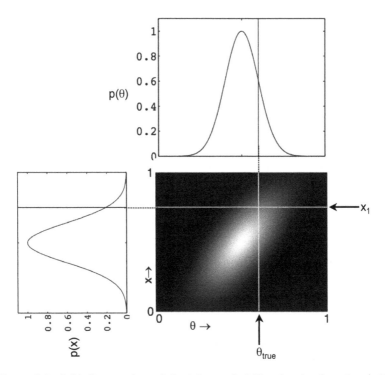

Figure 6.2: A bird's eye view of the joint probability density function (pdf) $p(X, \Theta)$ shown in Figure 6.1. The marginal pdfs of $p(X, \Theta)$ are the marginal likelihoods $p(X)$ (left) and the prior pdf $p(\Theta)$ (top).

Here, we have used the example of a population of coins with different biases, but this could be re-cast in terms of any number of similar scenarios. For instance, each coin bias θ could be replaced by the height of an individual chosen from a population, and the proportion x of observed heads for a coin could be replaced by the measured height of an individual, where this measurement includes some error in the form of measurement noise. Thus, rather than using a posterior pdf to estimate the bias of a single coin, given a noisy estimate of bias in the form of an observed proportion of heads, we would instead use a posterior pdf to estimate the height of one individual, given a noisy estimate of height in the form of an observed measurement of that individual's height.

6.2. A Bird's-Eye View of the Joint Distribution

A bird's-eye view of Figure 6.1 yields another representation of the joint probability distribution, shown in Figures 6.2 and 6.3a, where lighter grey-levels correspond to higher points on the surface in Figure 6.1.

Let's assume that the true (but unknown) value of our coin's bias is $\theta_{true} = 0.60$, as indicated by the vertical line in Figures 6.2, 6.3a. Let's also assume that we flip the coin N times to obtain an observed proportion of heads $x_1 = 0.75$. So, as usual, we have some noisy data x_1, which we will use to estimate the value of some parameter θ.

If we repeated these N coin flips many times then it can be shown that the distribution of x values has a Gaussian distribution, with a mean $x_{mean} = \theta_{true}$. The variability of x around x_{mean} is reflected in the width of the vertical cross-section at θ_{true}, which is also the width of the elliptical white region at θ_{true} in Figures 6.2,6.3a. The point is that the observed value x_1 of X given θ_{true} is a sample from a Gaussian distribution centred on x_{mean}, and the particular value we observe is the only information we have in order to estimate θ_{true}.

Using the joint distribution $p(X, \Theta)$, we can find that value of coin bias θ which is most consistent with x_1, the proportion of heads observed. A graphical way to do this is to draw a horizontal line across $p(X, \Theta)$ at $X = x_1$ (Figures 6.2,6.3a). This yields a cross-section with heights defined by $p(x_1, \Theta)$, where this cross-section is plotted in Figure 6.3b. The maximum value of the cross-section $p(x_1, \Theta)$ occurs at $\theta = 0.69$. This result is obtained by fixing the value of X at a given value x_1, and then finding the most probable value of Θ. In other words, this is the most probable value of Θ given x_1, and is therefore related to a posterior probability.

We can re-write $p(x_1, \Theta)$ in terms of the posterior distribution $p(\Theta|x_1)$ and the marginal likelihood $p(x_1)$

$$p(x_1, \Theta) \quad = \quad p(\Theta|x_1)p(x_1). \tag{6.1}$$

Re-arranging this shows that the posterior probability distribution is proportional to the cross-section of the joint distribution $p(x_1, \Theta)$ at x_1

$$p(\Theta|x_1) \quad = \quad p(x_1, \Theta)/p(x_1), \tag{6.2}$$

where the constant of proportionality is $1/p(x_1)$. We do not know the value of this constant, but as we are interested only in finding which value of θ maximises the posterior distribution, we do not need to know (and we can always work it out if needed, see p73). Crucially, this proportionality implies that the peak in the cross-section $p(x_1, \Theta)$ of the joint distribution occurs at exactly the same location as the peak in the posterior distribution $p(\Theta|x_1)$ (ie at $\theta = 0.69$). As the peak in the posterior distribution defines the maximum a posteriori (MAP) estimate θ_{MAP} of θ_{true}, it follows that the peak in the cross-section $p(x_1, \Theta)$ of the joint distribution also corresponds to the MAP estimate θ_{MAP}, so $\theta_{MAP} = 0.69$.

This is not intended to suggest that the joint distribution provides a short-cut past Bayes' rule, but to emphasize that Bayes' rule is really a method for estimating the true value of Θ without requiring access to the joint distribution (which we rarely have, in practice).

In order to reinforce this point, we will approach the problem from the more conventional end, where we have an observed value x_1 and we wish to estimate the true value θ_{true} using Bayes' rule.

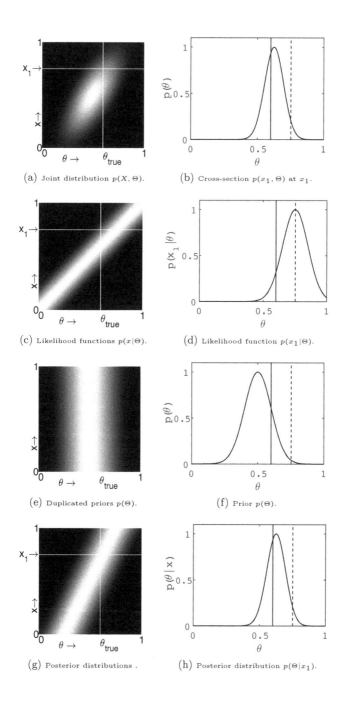

(a) Joint distribution $p(X, \Theta)$.

(b) Cross-section $p(x_1, \Theta)$ at x_1.

(c) Likelihood functions $p(x|\Theta)$.

(d) Likelihood function $p(x_1|\Theta)$.

(e) Duplicated priors $p(\Theta)$.

(f) Prior $p(\Theta)$.

(g) Posterior distributions .

(h) Posterior distribution $p(\Theta|x_1)$.

6.3. A Bird's-Eye View of Bayes' Rule

In the absence of a prior, we can use the observed value $x_1 = 0.75$ to plot the likelihood function $p(x_1|\Theta)$, shown in Figure 6.3d. This is a cross-section of the family of likelihood functions in Figure 6.3c, where each horizontal cross-section (row) represents the likelihood function for a different value x. The peak of the likelihood function is the maximum likelihood estimate (MLE) θ_{MLE} of Θ, and is indicated by the dashed vertical line in Figure 6.3d. In this case, we have $\theta_{MLE} = 0.75$.

In order to obtain the posterior distribution $p(\Theta|x_1)$, we need to multiply the likelihood function $p(x_1|\Theta)$ in Figure 6.3c by the prior distribution $p(\Theta)$ for Θ (and divide by the probability $p(x_1)$). In this example, the prior $p(\Theta)$ is one of the two marginal distributions of the joint distribution $p(X, \Theta)$ shown in Figures 6.1 and 6.3a, which is also plotted in Figure 6.3f.

Notice that each row of Figure 6.3c is a likelihood function for a different value of x, and that the posterior is obtained by multiplying this by the same prior distribution $p(\Theta)$ (and dividing by $p(x_1)$), which has been plotted in every row of Figure 6.3e. Therefore, multiplying each row in Figure 6.3e (the prior) by the likelihood function in the same row of Figure 6.3c (and dividing by $p(x_1)$) yields a family of

Figure 6.3: (Opposite) A bird's-eye view of Bayes' rule.
a) A bird's eye view of the 2D joint distribution $p(X, \Theta)$ shown in Figure 6.1. The horizontal line corresponds to the cross-section $p(x_1, \Theta)$ in (b), which is proportional to the posterior $p(\Theta|x_1)$ at $x_1 = 0.75$.
c) A family of likelihood functions. Each row (horizontal cross-section) defines a likelihood function $p(x|\Theta)$ for a specific value of X. The horizontal line corresponds to the cross-section shown in (f), which is the likelihood function $p(x_1|\Theta)$.
e) Each row is a copy of the marginal distribution $p(\Theta)$ shown in f.
g) Each row defines a posterior distribution $p(\Theta|x)$ for a different value of x. The horizontal line corresponds to the cross-section shown in h, which is the posterior distribution $p(\Theta|x_1)$.
The solid vertical lines correspond to the true value of $\theta_{true} = 0.6$, and the dashed vertical lines correspond to $\theta_{MLE} = 0.75$. Note that (b) and (h) have a maximum at the same value of θ. Graphs on the right have been scaled to have a maximum value of one.

```python
# Make prior  distribution p(THETA).
# Find likelihood function for each value of x.
# Make joint distribution p(X,THETA) and posteriors p(THETA|X).
# See online file for imports.
imax = 1001; imid = 500.0;
X = linspace(0,1,imax)
sdlik = 0.1                    # Standard deviation of likelihood.
sdprior = 0.1                  # Standard deviation of prior.
x1_index = round(imax*0.25)       # Index at x=0.75.
thetatrue_index = round(imax*0.6) # Index at theta=0.6.


# Make Gaussian prior.
prior =  np.exp(-((X-0.5)**2/(2*sdprior**2)))
# Show prior distribution.
ax6 = plt.subplot(426); plt.ylim([0, 1.1])
plt.plot(X,prior);       plt.title("Prior distribution")
plt.ylabel("$p(x)$");    plt.xlabel("$\\theta$")
plt.setp(ax6.get_xticklabels(), visible=False)
plt.setp(ax6.get_yticklabels(), visible=False)


likelihoods = zeros((imax,imax)) # Create 2D array for likelihoods.
prior2d = zeros((imax,imax)) # Create 2D array for duplicated priors.
# Fill arrays.
for r in arange(0,imax):
    rfrac = 1.0*(imax-r)/imax
    likelihoods[r,:] = np.exp(-((X-rfrac)**2/(2*sdprior**2)))
    prior2d[r,:] = prior

likelihoods = likelihoods/np.max(likelihoods)
prior2d = prior2d/np.max(prior2d) # Set array max value to 1.

lik1d = likelihoods[x1_index,:] # Get likelihood function at X=x1.
ax4 = plt.subplot(424);            plt.ylim([0,1.1])
plt.plot(X,lik1d);                 plt.title("Likelihood")
plt.ylabel("$p(x_1|\\theta)$");    plt.xlabel("$\\theta$")
plt.setp(ax4.get_xticklabels(), visible=False)
plt.setp(ax4.get_yticklabels(), visible=False)


# Find joint distribution: p(X,THETA)=p(X|THETA)*p(THETA), Eq 3.41.
joint1 = likelihoods * prior2d;    joint1=joint1/np.max(joint1)


# Get slice of joint distribution at X=x1.
joint1d = joint1[x1_index,:];    joint1d = joint1d/np.max(joint1d)
ax2 = plt.subplot(422);            plt.plot(X,joint1d)
plt.title("Joint distribution at $X=x_1$"); plt.ylim([0,1.1])
plt.ylabel("$p(x_1,\\theta)$");    plt.xlabel("$\\theta$")
plt.setp(ax2.get_xticklabels(),visible=False)
plt.setp(ax2.get_yticklabels(),visible=False)
```

```
# Find marginal likelihood p(X)
pX = np.sum(joint1,axis=1)
pX = pX/np.max(pX)

# Find posterior distribution for each value of x.
pX_duplicated = zeros((imax,imax))
for r in arange(0,imax):
    pX_duplicated[:,r]=pX
posteriors = likelihoods * prior2d / pX_duplicated
posteriors = posteriors/np.max(posteriors)

# Get posterior distribution at X=x1.
posterior = posteriors[x1_index,:]
ax8 = plt.subplot(428);         plt.ylim([0,1.1])
plt.plot(X,posterior)
plt.title("Posterior distribution")
plt.ylabel("$p(\\theta|x_1)$");    plt.xlabel("$\\theta$")
plt.setp(ax8.get_xticklabels(), visible=False)
plt.setp(ax8.get_yticklabels(), visible=False)

# Show 2D distributions as images.
lw = 5  # Determines white line width in 2d graphs.
# Show joint distribution.
# Mark all rows at thetatrue_index and columns at x1_index.
joint1[:,thetatrue_index-lw:thetatrue_index+lw]=1
joint1[x1_index-lw:x1_index+lw,:]=1
ax1 = plt.subplot(421);         plt.axis('off')
plt.imshow(joint1,cmap="gray"); plt.title("Joint distribution")

# Show likelihood functions.
plt.subplot(423)
# Mark all rows at thetatrue_index and columns at x1_index.
likelihoods[:,thetatrue_index-lw:thetatrue_index+lw]=1
likelihoods[x1_index-lw:x1_index+lw,:]=1
plt.imshow(likelihoods, cmap="gray")
plt.title("Likelihoods");     plt.axis('off')

# Show copies prior distribution.
plt.subplot(425);   plt.axis('off')
plt.imshow(prior2d,cmap="gray")
plt.title("Prior distributions")

# Show posterior distributions.
plt.subplot(427)
posteriors[:,thetatrue_index-lw:thetatrue_index+lw]=1
posteriors[x1_index-lw:x1_index+lw,:]=1
plt.imshow(posteriors, cmap="gray")
plt.title("Posterior distributions"); plt.axis('off');
figManager = plt.get_current_fig_manager()
figManager.window.showMaximized()
```

posterior distributions, each of which is proportional to the same row of the joint distribution in Figure 6.3a. In the case of the data x_1 we have observed, the posterior distribution is shown in Figure 6.3g, which has a maximum value at $\theta_{MAP} = 0.69$.

6.4. Slicing Through Joint Distributions

Here we explore the relationship between the joint distribution $p(X, \Theta)$, the likelihood function $p(x|\Theta)$ and the posterior distribution $p(\Theta|x)$.

Figure 6.3a shows that the coin bias θ_{true} of a coin defines a vertical line, which slices through the joint distribution at $\Theta = \theta_{true}$, and therefore defines a 1D distribution $p(X, \theta_{true})$. The peak of this distribution occurs at a value x which has no special name, but is the mean x_{mean} of the distribution $p(X, \theta_{true})$. In this example, $x_{mean} = \theta_{true}$ (and, in general, knowing x_{mean} determines θ_{true}).

When we flip a coin with bias θ_{true} a number of times, the proportion x_1 of heads observed is a noisy estimate of x_{mean}, and is therefore also a noisy estimate of θ_{true}. As we saw in Section 6.2, the observed value x_1 defines a horizontal line in Figure 6.3a. This line slices through the joint distribution at x_1, and therefore defines a 1D distribution $p(x_1, \Theta)$, which is proportional to the posterior probability distribution $p(\Theta|x_1)$. Therefore, the peak of the 1D distribution $p(x_1, \Theta)$ corresponds to the maximum a posteriori (MAP) estimate θ_{MAP}.

Crucially, if we do not have access to the joint distribution $p(X, \Theta)$ (and we rarely do) then we can use Bayes' rule to find exactly the same estimate θ_{MAP} as would be obtained if we did have access to the joint distribution. The reason that we do not usually have access to the joint distribution $p(X, \Theta)$ is discussed in the next section.

6.5. Statistical Independence

The notion of statistical independence is important for more advanced applications of Bayes' rule. In the examples above, the variables X and Θ are correlated, so that a knowing the value of X tells us something about the probable value of Θ, and vice versa. However, if X and Θ

were statistically independent then the value of X tells us absolutely nothing about the value of Θ, and vice versa (see Appendix C).

Formally, if X and Θ are independent then their joint 2D probability density function $p(X, \Theta)$ is exactly the same as the product of its two marginal probability density functions $p(X)$ and $p(\Theta)$ (shown on the back-planes of Figure 6.1). Thus, $p(X, \Theta)$ factorises into two 1D probability density functions, $p(X)$ and $p(\Theta)$ so that

$$p(X, \Theta) = p(X) \times p(\Theta). \tag{6.3}$$

If X and Θ are correlated, as in Figure 6.1, then they are not independent, so their joint probability density function $p(X, \Theta)$ does not factorise, so that

$$p(X, \Theta) \neq p(X) \times p(\Theta). \tag{6.4}$$

The reason that this matters is that, if k variables are independent then it is possible to represent their joint probability density function as k 1D probability density functions, rather than a single k-dimensional distribution. As k increases, k 1D probability density functions represent an increasingly large saving over a single k−dimensional joint probability density function in terms of the amount of storage space (computer or biological) and the time required to search this space.

For example, consider the space required to store the joint probability density function shown in Figure 6.1, for which $k = 2$. If $N = 10$ intervals along each axis (eg $0 - 0.1, \ldots, 0.9 - 1.0$) then this defines

$$S_{joint} = 10^2, \tag{6.5}$$

or 100 checker-board squares on the ground plane, and the density in each square must be represented. But if $p(X)$ and $p(\Theta)$ were independent then we would need to store $N = 10$ density values along each axis, making

$$S_{factorised} = 2 \times 10, \tag{6.6}$$

or 20 values that must be stored. Thus, even in this fairly minimal example, the amount of storage required to represent a 2D joint probability density function is 100 density values if the variables are not independent, but only 20 density values if they are independent.

Because each new variable adds one spatial dimension to the probability density function, which multiplies the total storage required by a constant factor, this is known as *the curse of dimensionality.*

Summary

When we observe a proportion x of heads, we now know that this is only half the story. The other half is implicit in the statistical co-occurrence of coin biases Θ and X values, which define the structure of the joint distribution $p(X, \Theta)$. But we do not have access to this joint distribution. If we did, then we could simply 'look up' the bias that most commonly co-occurs with the proportion x of heads we observed, by plotting the cross-section $p(x, \Theta)$ of the joint distribution at x.

As we do not have access to the joint distribution $p(X, \Theta)$, we are forced to rely on the posterior distribution $p(\Theta|x)$, which is proportional to the cross-section $p(x, \Theta)$ of the joint distribution $p(X, \Theta)$ at $X = x$. Fortunately, Bayes' rule provides this posterior distribution by combining the likelihood with the prior. So, in a sense, Bayes' rule is what we are forced to use if we do not have access to the joint distribution.

Chapter 7

Bayesian Wars

> Far better an approximate answer to the right question ...
> than an exact answer to the wrong question.
> John Tukey, 1962.

Introduction

To some people, the word *Bayesian* is like a red rag to a bull. Such strong reactions are the result of many years of controversy and misunderstanding regarding the nature of probability. We will not dwell on the history of this controversy here, but we will try to uncover what it is about, and why it still rages.

7.1. The Nature of Probability

There are essentially two types of statisticians, *frequentists* and *Bayesians*, and there is usually little love lost between them. The main difference is that frequentists consider probability to be a property of the physical world, whereas Bayesians consider probability to be a measure of uncertainty regarding their knowledge of the physical world. This subtle difference appears quite trivial, but it has far-reaching consequences.

Probability as Information

Consider what happens in the throw of a die. Before it lands, you know the probability that it shows the number 3 is 1/6, or about 0.17. After it lands, and with the benefit of hindsight, it is apparent that the probability of a 3 was actually 1/1 or 1.0. So if we could replay every aspect of your throw exactly then it would again show a 3. So what was the probability of observing a 3, was it 0.17 or 1?

This question lies at the heart of the concept of probability. To a frequentist, probability represents the proportion of times a specific outcome occurs (eg a 3). To a Bayesian, probability is a measure of the amount of certainty that an observer has about a specific outcome.

Before a die is thrown, we have so little information that the only rational way to proceed is to assume all outcomes are equally probable. However, if we could slow down time and watch the die in the second before it stops then it may be balanced on one edge. At that precise point in time, we would state that there are only two possible outcomes, say 3 and 6. As the die tips towards the 3-face we would state that there is only one possible outcome. In summary, as the amount of information we have increases, our confidence in the probability of each possible outcome also increases. This suggests that probability is not a property of the physical world, but is a measure of how much information an observer has about that world.

Dicing with 100 balls: Imagine that you are presented with an opaque bin containing 50 white balls and 50 black balls, as in Figure 7.1a. You remove one ball, labelled ball A, but you do not look at it, so it remains hidden, as in Figure 7.1b. What is your estimate of the probability that ball A is black? If we define θ_b to be the probability that ball A is black then your answer should be $p(\theta_b|I_1) = 0.5$, where I_1 is any relevant information (eg there are equal numbers of black and white balls). Now I remove one ball, which we will label ball B. I show you ball B, which is white, as in Figure 7.1c. Given this new information, which we can label as I_2, what is your estimate of the probability that ball A is black? The precise answer does not matter. What does matter is that your estimate of the probability that ball A is black should have changed.

In order to explore this, let's replay this scenario in more detail. You estimated $p(\theta_b|I_1)$ to be 0.5 at the time you chose ball A. But you have now received new information I_2, which allows you to update your

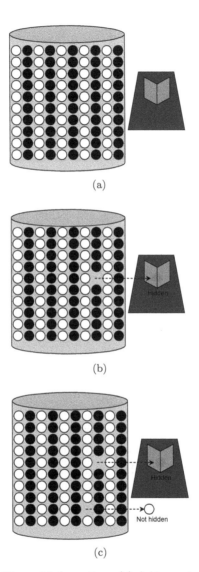

(a)

(b)

(c)

Figure 7.1: Probability and information. (a) A bin contains 50 white and 50 black balls. (b) A ball, labelled ball A, is chosen at random and kept hidden from sight. What is the probability that ball A is black? (Answer=0.5) (c) A second ball, labelled ball B, is then chosen at random, and is found to be white. Given this new information, what is the probability that ball A is black? (Answer > 0.5)

estimate of $p(\theta_b|I_1)$. In other words, your estimate of the probability that ball A is black should change when you learn that ball B is white.

In order to assess precisely how your estimate of $p(\theta_b)$ should change, let's consider the consequences of ball A being black, and then white. If ball A is black then there must have been 49 black and 50 white balls when I chose ball B. So the probability that I chose a black ball B was 49/99 (ie just under 0.5), whereas the probability that I chose a white ball B was 50/99 (ie just over 0.5). Conversely, if ball A is white then there must have been 49 white and 50 black balls when I chose ball B. In this case, the probability that I chose a white ball B was 49/99 (ie just under 0.5), whereas the probability that I chose a black ball B was 50/99 (ie just over 0.5).

Now if I show you that ball B is white then this implies that there were probably more white balls than black when I chose my ball. This, in turn, implies that the ball A you chose was more probably black than white, $p(\theta_b|I_2) > 0.5$. This is true even though you knew there was a 50/50 chance of choosing a black ball when you made your choice.

Notice that throughout this scenario, the colour of ball A has not been disclosed. Indeed, it could be black or white, with probabilities that differ by about one part in 100 (ie by about 0.01). The important point is not what colour ball A has, but how your estimate of the probability of ball A being black or white is altered by new information.

Indeed, as we have seen, the final decision regarding the colour of ball A has almost (but not quite) a 50% chance of being wrong. The point is that the best decision that could have been made, based on the available information, was in fact made. Even if such a decision led to the wrong answer, it was still the rational thing to do. This point was eloquently expressed in around 500 BC by Herodotus[38] (possibly the first Bayesian):

> A decision was wise, even though it led to disastrous consequences, if the evidence at hand indicated it was the best one to make; and a decision was foolish, even though it led to the happiest possible consequences, if it was unreasonable to expect those consequences.

Dicing with two balls: If the above example appears opaque then a stripped-down version of it should be more transparent. As above, you are presented with a bin, but this bin contains exactly one black and one white ball, as in Figure 7.2a. You remove one ball, which we again call ball A, but you do not look at it, as in Figure 7.2b. What is your estimate of the probability that ball A is black? Clearly, the answer is $p(\theta_b|I_1) = 0.5$, where I_1 represents any relevant information available (eg there is one black and one white ball present).

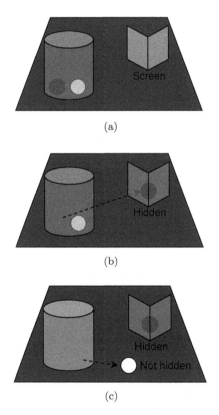

Figure 7.2: Probability and information. **a)** A bin contains one white and one black ball. **b)** One ball A is chosen at random and kept hidden from sight. What is the probability that ball A is black? (Answer=0.5). **c)** The second ball is then removed and is found to be white. Now, what is the probability that ball A is black? (Answer=1.0).

Now, I remove the other ball, which we call ball B. Its colour is white, as in Figure 7.2c. Again, we can ask: what is your estimate of the probability that ball A is black? It is now apparent that the ball A you chose must be black, so that $p(\theta_b|I_2) = 1.0$.

This is not, as it may appear, a contradiction. The first figure $(p(\theta_b|I_1) = 0.5)$ is your estimate of the probability that ball A is black when you knew only that there was one black and one white ball present. In contrast, the second figure $(p(\theta_b|I_2) = 1)$ is your estimate of the probability that ball A is black after you received new information in the form of the colour of ball B. Both estimated probabilities are correct, because they are each based on different amounts of information, I_1 and I_2.

Note that this example is logically identical to the example above which used 100 balls, but the consequences of new information are more obvious and more compelling in this two-ball case. However, in both cases, the general lesson is the same.

If probability is a measure of the state of the physical world then the probability that the ball A you chose is black should not change over time (eg after you learn that the other ball is white). On the other hand, if probability is a measure of how much information an observer has about the state of the physical world then it should change when new information becomes available.

The above examples demonstrate that, if an observer is rational then their estimate of probability does change when new information becomes available. This implies that probability is something an observer attributes to the physical world, rather than being a property of the world. In essence, probability is a measure of how much information an observer has about the state of the physical world.

Subjective Probability

Just as one individual should alter his estimate of probability when new information becomes available, so two different individuals should have different estimates of a probability if these estimates are based on different information.

As an example, if two individuals observe different numbers of coin flips of the same coin then their estimates of the posterior probability that the next coin flip will yield a head will usually differ. In such cases, each individual is said to have a different *subjective probability* for the next flip outcome. Specifically, if person A observes 10 coin flips (7 heads followed by 3 tails), but person B is only shown the final 4 coin flips (1 head and 3 tails) then A and B have different information. Accordingly, their estimates of the coin's bias will be different, and therefore their predictions regarding the outcome of the next flip will also differ. Specifically, if both observers have a uniform prior for coin bias then A has a MAP estimate of the coin's bias of 0.7, whereas B has a MAP estimate of 0.25. By definition, each of these estimated bias values represents one individual's estimate of the probability that the next coin flip will yield a head. Clearly, there is nothing subjective about these two probabilities, if the term "subjective" is taken to mean they are based on personal opinion. They are simply two posterior probabilities, but based on different amounts of information. As we saw in the previous section, probability can be interpreted in terms of the amount of information we have about some aspect of the physical world. The fact that this has been labelled as "subjective probability" is unfortunate because it gives the false impression that Bayesian analysis depends on personal opinions (see Section 4.8, p99).

7.2. Bayesian Wars

Bayes' rule is not just a convenient rule which happens to work. It is part of a set of rules which are logically implied by our insistence that such rules should yield results which conform to the behaviour of the physical world. Whereas engineers and physicists view Bayes' rule as plain common sense, others view Bayes' rule with deep suspicion. Consequently, the on-going skirmishes and battles[20] that surround the debate about Bayes' rule occasionally break out into a Bayesian war.

One of the key players in the development of modern Bayesian theory, Jaynes[18], pointed out that much of the heat of this debate depends on ideological arguments regarding the interpretation of probability. Specifically, (as stated in Chapter 1) in setting up rules for calculating with probabilities, we would insist that such calculations should be consistent with our everyday experience of the physical world, just as surely as we would insist that $1 + 1 = 2$. Indeed, Cox(1946)[7] and Kolmogorov(1933)[24] independently provided mathematical frameworks for probability, both of which yield Bayes' rule.

Bayes' rule has been used to combine evidence from different sources in a wide variety of different applications, such as medical diagnosis, computer vision, and signal processing. Of course, a court of law is where evidence from different sources must be weighed carefully. However, in the 1996 British rape case of Adams, despite valiant efforts to teach the basics of Bayesian reasoning to the jury, the Appeals Court concluded that the use of Bayes' rule is, "a recipe for confusion and misjudgment, possibly even among counsel, but very probably among judges and, as we conclude, almost certainly among jurors"[9].

Such dogmatic, yet unsurprising, declarations should ensure that we are absolutely clear about the logical status of Bayes' rule. Bayes' rule is not like a 'rule of thumb' which was discovered in the process of observing how probabilities interact. It is not a hypothesis awaiting empirical tests of its validity, like 'red sky at night, shepherd's delight'. Formally, Bayes' rule is a *theorem*, which means it has been proven to be true. So why does it generate so much emotion?

Much of the criticism regarding Bayes' rule involves the question of priors. In essence, dissenters claim that choosing a prior amounts to choosing a particular class of prejudices regarding some observed data, and that the conclusions of Bayesian inference are therefore tainted by those prejudices. The counter-arguments rely on the following points.

First, a non-Bayesian analysis is usually based on some form of maximum likelihood estimation, which merely ignores the question of which prior to use, and implicitly employs a uniform prior. Second, this implicit uniform prior can lead to erronous results because a non-uniform reference prior may be required (see Section 4.8). Third, given sufficiently large amounts of data, the impact of the prior on any conclusion is minimal because, as the amount of data increases, so the influence of the prior on the posterior usually diminishes (see Chapter 4). Thus, the results from Bayesian and non-Bayesian methods (strictly, those based on likelihoods) usually converge as the amount of data increases. However, as the rate and degree of convergence can only be assessed by comparing the results of a non-Bayesian analysis with a Bayesian analysis, the decision as to which analysis to choose seems self-evident. Moreover, to take a realistic example, the eye's photoreceptors are light-starved even in daylight conditions[25], suggesting there is insufficient data under natural conditions.

Beyond these issues, there are differences between Bayesian analysis and frequentist methods which seem to have an almost philosophical flavour. However, this introductory text cannot do justice to the vast quantities of text written about these differences. As discussed above, within the Bayesian framework, all that we can ever know about a given parameter is based on noisy measurements, which can be used to infer a posterior distribution for that parameter. Thus, a Bayesian analysis is based on the assumption that a parameter is a random variable, and our knowledge of it takes the form of a posterior distribution that can be estimated from an observed set of data. This posterior distribution can be summarised by its peak, which defines the maximum a posteriori (MAP) estimate, and a width, which defines the degree of certainty associated with the MAP estimate. In contrast, a frequentist perspective treats each parameter as a fixed property of the physical world, which can be estimated using (for example) the mean of a large number of noisy measurements.

Of course, a pre-requisite for estimating parameter values is the specification of a model for the data under consideration. This model may be as simple as a Gaussian distribution (with parameters μ and σ), linear regression (with parameters m and c), or more complex models. The point is that a Bayesian analysis entails some form of model-fitting, which forces a consideration of the prior distributions for the various parameters in the model. Thus, leaving aside the mechanics of Bayesian analysis, the process of specifying a model for a given data set forces us to make explicit all of our assumptions regarding the process of fitting a model. A fortuitous side-effect is that we *always write down the probability of everything.* (Steve Gull, quoted from MacKay[28] (p61)).

7.3. A Very Short History of Bayes' Rule

Bayes' rule probably began life in the parsonage of an 18th century preacher and amateur mathematician, the Reverend Thomas Bayes (1702-1761)[39], who lived in Tunbridge Wells, in the county of Kent, England. He published only one mathematical paper in his lifetime, on Newton's calculus or *fluxions* as it was then called. However, his main achievement was a paper published in 1763 after his death, "An essay towards solving a problem in the doctrine of chances"[1], which was sent to the Royal Society of London by his friend Richard Price. The blind Cambridge mathematician Nicholas Saunderson (1682-1739)[39] is also considered by some to have discovered Bayes' theorem.

The French mathematician Pierre-Simon Laplace (1749-1827) independently discovered Bayes' rule in 1812. For this reason, Bayes' rule is sometimes called the *Bayes-Laplace rule.* Despite Laplace's success in applying his methods to a variety of problems, including the mass of Jupiter, and medical statistics, Bayesian methods were eclipsed in the 19th and 20th centuries by non-Bayesian (frequentist) methods. However, in the 20th century, Bayes' approach was championed first by Jeffreys (1939)[19], and then by Jaynes (2003)[18]. Recently, texts which demonstrate the utility of Bayesian methods have helped increase the dissemination of Bayesian methods[14;27;38].

Summary

According to the physicist, Richard Feynman,

> Science is a way of trying not to fool yourself. The first principle is that you must not fool yourself, and you are the easiest person to fool. (Richard Feynman, 1974).

In essence, Bayes' rule provides a method for not fooling ourselves into believing our own prejudices, because it represents a rational basis for believing things that are probably true, and for disbelieving things that are probably not.

Further Reading

Bernardo JM and Smith A (2000)[4]. Bayesian Theory. *A rigorous account of Bayesian methods, with many real-world examples.*

Bishop C (2006)[5]. Pattern Recognition and Machine Learning. *As the title suggests, this is mainly about machine learning, but it provides a lucid and comprehensive account of Bayesian methods.*

Cowan G (1998)[6]. Statistical Data Analysis. *An excellent non-Bayesian introduction to statistical analysis.*

Dienes Z (2008)[8]. Understanding Psychology as a Science: An Introduction to Scientific and Statistical Inference. *Provides tutorial material on Bayes' rule and a lucid analysis of the distinction between Bayesian and frequentist statistics.*

Gelman A, Carlin J, Stern H and Rubin D (2003)[14]. Bayesian Data Analysis. *A rigorous and comprehensive account of Bayesian analysis, with many real-world examples.*

Jaynes E and Bretthorst G (2003)[18]. Probability Theory: The Logic of Science. *The modern classic of Bayesian analysis. It is comprehensive and wise. Its discursive style makes it long (600 pages) but never dull, and it is packed full of insights.*

Khan S (2012). Conditional probability with Bayes Theorem.
Salman Khan's online mathematics videos make a good introduction to various topics, including Bayes' rule.
www.khanacademy.org

Lee PM (2004)[27]. Bayesian Statistics: An Introduction. *A rigorous and comprehensive text with a strident Bayesian style.*

MacKay DJC (2003)[28]. Information theory, inference, and learning algorithms. *The modern classic on information theory. A very readable text that roams far and wide over many topics, almost all of which make use of Bayes' rule.*

Migon HS and Gamerman D (1999)[30]. Statistical Inference: An Integrated Approach. *A straightforward (and clearly laid out) account of inference, which compares Bayesian and non-Bayesian approaches. Despite being fairly advanced, the writing style is tutorial in nature.*

Pierce JR (1980)[34], 2nd Edition. An introduction to information theory: symbols, signals and noise. *Pierce writes with an informal, tutorial style of writing, but does not flinch from presenting the fundamental theorems of information theory.*

Reza FM (1961)[35]. An introduction to information theory. *A more comprehensive and mathematically rigorous book than the Pierce book above, and should ideally be read only after first reading Pierce's more informal text.*

Sivia DS and Skilling J (2006)[38]. Data Analysis: A Bayesian Tutorial. *This is an excellent tutorial style introduction to Bayesian methods.*

Spiegelhalter D and Rice K (2009)[36]. Bayesian statistics. Scholarpedia, 4(8):5230. `www.scholarpedia.org/article/Bayesian_statistics`
A reliable and comprehensive summary of the current status of Bayesian statistics.

Appendices

Appendix A

Glossary

Bayes' rule Given some observed data x, the posterior probability that the parameter Θ has the value θ is $p(\theta|x) = p(x|\theta)p(\theta)/p(x)$, where $p(x|\theta)$ is the likelihood, $p(\theta)$ is the prior probability of the value θ, and $p(x)$ is the marginal probability of the value x.

conditional probability The probability that the value of one random variable Θ has the value θ given that the value of another random variable X has the value x; written as $p(\Theta = \theta|X = x)$ or $p(\theta|x)$.

forward probability Reasoning forwards from the known value of a parameter to the probability of some event defines the forward probability of that event. For example, if a coin has a bias of θ then the forward probability $p(x_h|\theta)$ of observing a head x_h is θ.

independence If two variables X and Θ are independent then the value x of X provides no information regarding the value θ of the other variable Θ, and vice versa.

inverse probability
Reasoning backwards from an observed measurement x_h (eg coin flip) involves finding the posterior or inverse probability $p(\theta|x_h)$ of an unobserved parameter θ (eg coin bias).

joint probability The probability that two or more quantities simultaneously adopt specified values. For example, the probability that a coin flip yields a head x_h and that a (possibly different) coin has a bias θ is the joint probability $p(x_h, \theta)$.

likelihood The conditional probability $p(x|\theta)$ that the observed data X has the value x given a putative parameter value θ is the likelihood of θ, and is often written as $L(\theta|x)$. When considered over all values Θ of θ, $p(x|\Theta)$ defines a *likelihood function*.

marginal distribution A distribution resulting from marginalisation of a multivariate (eg 2D) distribution. For example, the 2D distribution $p(X, \Theta)$ shown in Figure 3.4 has two marginal distributions, which, in this case, are the prior distribution $p(\Theta)$ and the distribution of marginal likelihoods $p(X)$.

maximum a posteriori (MAP) Given some observed data x, the value θ_{MAP} of an unknown parameter Θ that makes the posterior probability $p(\Theta|x)$ as large as possible is the maximum a posteriori or MAP estimate of the true value θ_{true} of that parameter. The MAP takes into account both the current evidence in the form of x as well as previous knowledge in the form of prior probabilities $p(\Theta)$ regarding each value of θ.

maximum likelihood estimate (MLE) Given some observed data x, the value θ_{MLE} of an unknown parameter Θ that makes the likelihood function $p(x|\Theta)$ as large as possible is the maximum likelihood estimate (MLE) of the true value of that parameter.

noise Usually considered to be the random jitter that is part of a measured quantity.

non-informative prior See reference prior, and Section 4.8.

parameter A variable (often a random variable), which is part of an equation which, in turn, acts as a model for observed data.

posterior The posterior probability $p(\theta|x)$ is the probability that a parameter Θ has the value θ, based on current evidence (data, x) and prior knowledge. When considered over all values of θ, it refers to the posterior probability distribution $p(\Theta|x)$.

prior The prior probability $p(\theta)$ is the probability that the random variable Θ adopts the value θ. When considered over all values Θ, it is the prior probability distribution $p(\Theta)$.

probability There are many definitions of probability. The two main ones are (using coin bias as an example): 1) Bayesian probability: an observer's estimate of the probability that a coin will land heads up is based on all the information the observer has, including the proportion of times it was observed to land heads up in the past. 2) Frequentist probability: the probability that a coin will land heads up is given by the proportion of times it lands heads up, when measured over a large number of coin flips.

probability density function (pdf) The function $p(\Theta)$ of a continuous random variable Θ defines the probability density of each possible value of Θ. The probability that $\Theta = \theta$ can be considered as the probability density $p(\theta)$ (it is actually the product $p(\theta) \times d\theta$).

probability distribution The distribution of probabilities of different values of a variable. The probability distribution of a continuous variable is a *probability density function*, and the probability distribution of a discrete variable is a *probability function*. When we refer to a case which includes either continuous or discrete variables, we use the term *probability distribution* in this text.

probability function (pf) A function $p(\Theta)$ of a discrete random variable Θ defines the probability of each possible value of Θ. The probability that $\Theta = \theta$ is $p(\Theta = \theta)$ or more succinctly $p(\theta)$. This is called a *probability mass function* (pmf) in some texts.

product rule The joint probability $p(x, \theta)$ is given by the product of the conditional probability $p(x|\theta)$ and the probability $p(\theta)$; that is, $p(x, \theta) = p(x|\theta)p(\theta)$. See Appendix C.

random variable (RV) Each value of a random variable can be considered as one possible outcome of an experiment that has a number of different possible outcomes, such as the throw of a die. The set of possible outcomes is the sample space of a random variable. A discrete random variable has a probability function (pf), which assigns a probability to each possible value. A continuous random variable has a probability density function (pdf), which assigns a probability density to each possible value. Upper case letters (eg X) refer to random variables, and (depending on context) to the set of all possible values of that variable. See Section 2.1, p31.

real number A number that can have any value corresponding to the length of a continuous line.

regression A technique used to fit a parametric curve (eg a straight line) to a set of data points.

reference prior A prior distribution that is 'fair'. See Section 4.8, p99 and Appendix H, p169.

standard deviation The standard deviation of a variable is a measure of how 'spread out' its values are. If we have a sample of n values of a variable x then the standard deviation of our sample is

$$\sigma = \sqrt{\frac{1}{n} \sum_{i=1}^{n} (x_i - \bar{x})^2}, \tag{A.1}$$

where $\bar{x}$ is the mean of our sample. The sample's variance is σ^2.

sum rule This states that the probability $p(x)$ that $X = x$ is the sum of joint probabilities $p(x, \Theta)$, where this sum is taken over all N possible values of Θ,

$$p(x) = \sum_{i=1}^{N} p(x, \theta_i).$$

Also known as the law of total probability. See Appendix C.

variable A variable is essentially a 'container', usually for one number. We use the lower case (eg x) to refer to a particular value of a variable.

Appendix B

Mathematical Symbols

$\propto$ indicates *proportional to* (similar to the symbol for infinity ∞).

$\sum$ the capital Greek letter sigma represents summation. For example, if we represent the $N = 3$ numbers 2, 5 and 7 as $x_1 = 2, x_2 = 5, x_3 = 7$ then their sum S is

$$S = \sum_{i=1}^{N} x_i \tag{B.1}$$

$$= x_1 + x_2 + x_3 \tag{B.2}$$

$$= 2 + 5 + 7 \tag{B.3}$$

$$= 14.$$

The variable i is counted up from 1 to N, and, for each i, the term x_i adopts a new value, and is added to a running total.

$\prod$ the capital Greek letter pi represents multiplication. For example, if we use the values defined above then the product of these $N = 3$ integers is

$$P = \prod_{i=1}^{N} x_i \tag{B.4}$$

$$= x_1 \times x_2 \times x_3 \tag{B.5}$$

$$= 2 \times 5 \times 7 \tag{B.6}$$

$$= 70. \tag{B.7}$$

The variable i is counted up from 1 to N, and, for each i, the term x_i adopts a new value, and is multiplied by a running total.

$\approx$ means 'approximately equal to'.

μ Greek letter mu (pronounced mew) is the mean of a variable.

η Greek letter eta (pronounced eater).

σ Greek letter sigma is the standard deviation of a distribution.

E mean or *expected value* of a variable (eg E[x]).

N the number of observations in a data set (eg coin flip outcomes).

Θ (capital theta) the set of values in the range θ_{min} to θ_{max}. If Θ is a random variable then the probability that Θ adopts a specific value θ is the value of probability distribution $p(\Theta)$ at $\Theta = \theta$.

θ Greek letter theta, a value of the random variable Θ.

X the set of values in the range x_{min} to x_{max}. If X is a random variable then the probability that X adopts a specific value x is defined by the value of the probability distribution $p(X)$ at $X = x$.

x a value of the random variable X.

$\mathbf{x}$ permutation (round brackets $(x_1, \ldots)$) or a combination (curly brackets $\{x_1, \ldots\}$) of x values.

x^n x raised to the power n. If $x = 2$ and $n = 3$ then $x^n = 2^3 = 8$.

$x!$ pronounced x *factorial*. For example, if $x = 3$ then $x! = 3 \times 2 \times 1 = 6$.

$p(X)$ the probability distribution of the random variable X.

$p(x)$ the probability that the random variable X has the value x.

$p(\Theta)$ the probability distribution of the random variable Θ.

$p(\theta)$ the probability that the random variable Θ has the value θ.

$p(X, \Theta)$ the joint probability distribution of the random variables X and Θ. For discrete variables, this is called the *joint probability function* (pf) of X and Θ, and for continuous variables, it is called the *joint probability density function* (pdf) of X and Θ.

$p(x, \theta)$ the *joint probability* that the random variables X and Θ have the values x and θ, respectively.

$p(x|\theta)$ the conditional probability that the random variable $X = x$ given that $\Theta = \theta$ is the *likelihood* of θ.

$p(x|\Theta)$ the set of conditional probabilities that $x = X$ given each possible value of Θ is the *likelihood function* of Θ.

$p(\theta|x)$ the conditional probability that the random variable $\Theta = \theta$ given that $X = x$ is the *posterior probability* that $\Theta = \theta$.

$p(\Theta|x)$ the set of conditional probabilities of each value of the random variable Θ given that $X = x$ is the *posterior probability distribution* of Θ.

() Round brackets. By convention, these are used to indicate the argument of a function (eg $y = f(x)$). Also used to denote an ordered set of items, such as a permutation or sequence (eg $\mathbf{x} = (1, 2, 3)$).

{} Curly brackets. By convention, these are used to denote an unordered set of items, such as a combination or the values (sample space) that a random variable can adopt (eg $X = \{1, 2, 5\}$).

[] Square brackets. In this book, these are used in place of round brackets when the latter make reading text difficult (eg $p(X) = [p(x_1), p(x_2), p(x_3)]$).

Appendix C

The Rules of Probability

We assume the letter x represents some observed data, and the Greek letter θ (theta) represents the value of a parameter Θ, and we wish to estimate θ by making use of the data x.

Independent Outcomes: If a set of individual outcomes are independent then the probability of that outcome set is obtained by multiplying the probabilities of each of the individual outcomes together.

For example, consider a coin for which the probability of a head x_h is $p(x_h) = 0.9$, and the probability of a tail x_t is therefore $p(x_t) = (1 - 0.9) = 0.1$. If we flip this coin twice then there are 4 possible pairs of outcomes: two heads (x_h, x_h), two tails (x_t, x_t), a head followed by a tail (x_h, x_t), and a tail followed by a head (x_t, x_h). In order to work out some averages, imagine that we flip this coin twice, 100 times. We label each flip according to whether it came first or second within a pair of flips, so we have 100 *first flip* outcomes, and 100 corresponding *second flip* outcomes (see Table C.1).

	h	t	$\{h, h\}$	$\{t, t\}$	(h, t)	(t, h)	$\{t, h\}$
N	90	10	81	1	9	9	18
$N/100$	0.90	0.10	0.81	0.01	0.09	0.09	0.18

Table C.1: The number N and probability $N/100$ of each possible outcome from 100 pairs of coin flips of a coin which lands heads up 90% of the time. Ordered sequences or permutations are written in round brackets '()', whereas un-ordered sets or combinations are written in curly brackets '{}'.

Given that $p(x_h) = 0.9$, we expect 90 heads and 10 tails within the set of 100 first flips, and the same applies to the set of 100 second flips. But what about the number of pairs of outcomes?

For each head obtained on the first flip, we can observe the corresponding outcome on the second flip, and then add up the number of pairs of each type (eg x_h, x_h). We already know that there are (on average)

$$90 = 0.9 \times 100, \tag{C.1}$$

heads within the set of 100 first flip outcomes. For each of these 90 heads, the outcome of each of the corresponding 90 second flips does not depend on of the outcome of the first flip, so we would expect

$$81 = 0.9 \times 90, \tag{C.2}$$

of these 90 second flip outcomes to be heads. In other words, 81 out of 100 pairs of coin flips should yield two heads. Notice that the figure of 90 heads was obtained from Equation C.1, so we can re-write Equation C.2 as

$$81 \quad = \quad 0.9 \times (0.9 \times 100) \tag{C.3}$$
$$= \quad 0.81 \times 100, \tag{C.4}$$

where 0.9 is the probability $p(x_h)$ of a head, so the probability of obtaining two heads is $p(x_h)^2 = 0.9^2 = 0.81$.

A similar logic can be applied to find the probability of the other pairs (x_h, x_t) and (x_t, x_t). For the pair (x_t, x_t), there are (on average) 10 tails observed in the set of 100 first flip outcomes. For each of these 10 flips, each of the corresponding 10 second flips also has an outcome, and we would expect $1 = 0.1 \times 10$ of these to be a tail too; so that 1 out of 100 pairs of coin flips should consist of two tails (x_t, x_t).

The final pair is a little more tricky, but only a little. For the ordered pair (x_h, x_t), as there are (on average) 90 heads from the set of 100 first flips, and we would expect $9 = 0.1 \times 90$ of the corresponding 90 second flips to yield a tail; so that 9 out of 100 pairs of coin flips should be (x_h, x_t) tails. Similarly, for the ordered pair (x_t, x_h), as there are (on average) 10 tails in the set of 100 first flips, we would expect $9 = 0.9 \times 10$ of the corresponding 10 second flips to yield a head; so that 9 out of 100 pairs of coin flips should be (x_t, x_h). If we now consider the number of pairs that contain a head and a tail *in any order* then we would expect there to be $18 = 9 + 9$ pairs that contain a head and a tail. Notice that the figure of 90 heads was obtained from $90 = 0.9 \times 100$, so we can write this as $9 = (0.1 \times 0.9) \times 100$, or $p(x_h)p(x_t) \times 100$.

In summary, given a coin that lands heads up on 90% of flips, in any given pair of coin flips we have (without actually flipping a single coin) worked out that there is an 0.81 probability of obtaining two heads, an 0.01 probability of obtaining two tails, and an 0.18 probability of obtaining a head and a tail. Notice that these three probabilities sum to one, as they should. More importantly, the probability of obtaining each pair of outcomes is obtained by multiplying the probability associated with each individual coin flip outcome.

We can apply this rule to any number of coin flips. For example, consider a coin flipped 5 times in succession, which yields the outcome sequence $(x_h, x_h, x_t, x_t, x_h)$. Given that these individual outcomes are independent, the probability of the observed sequence is

$$
\begin{aligned}
p((x_h, x_h, x_t, x_t, x_h)) &= p(x_h)p(x_h)p(x_t)p(x_t)p(x_h) & \text{(C.5)} \\
&= p(x_h)^3 \times p(x_t)^2 & \text{(C.6)} \\
&= 0.007. & \text{(C.7)}
\end{aligned}
$$

Note that this is the probability of a particular sequence or permutation, and the probability of 3 heads and 2 tails in any order involves a constant in the form of a binomial coefficient (see Appendix E).

Conditional Probability: The conditional probability that $X = x$ given that $\Theta = \theta$ is defined as

$$
p(x|\theta) = p(x, \theta)/p(\theta). \tag{C.8}
$$

The Product Rule: Multiplying both sides of Equation C.8 by $p(\theta)$ yields the *product rule* (also known as the *chain rule*)

$$p(x, \theta) \quad = \quad p(x|\theta)p(\theta). \tag{C.9}$$

The Sum Rule and Marginalisation: The *sum rule* is also known as the law of total probability. In the case of a discrete variable,

$$p(x) \quad = \quad \sum_{i=1} p(x, \theta_i), \tag{C.10}$$

and, applying the product rule, yields

$$p(x) \quad = \quad \sum_{i=1} p(x|\theta_i)p(\theta_i). \tag{C.11}$$

In the case of a continuous variable, the sum and product rules yield

$$p(x) = \int_\theta p(x, \theta)\, d\theta \quad = \quad \int_\theta p(x|\theta)p(\theta)\, d\theta. \tag{C.12}$$

This is known as *marginalisation*, and yields the marginal probability $p(x)$ of the joint distribution $p(X, \Theta)$ at $X = x$. When applied to all values of X, marginalisation yields the marginal probability distribution $p(X)$ of $p(X, \Theta)$.

Bayes' Rule: If we swap θ for x in Equation C.9 then

$$p(\theta, x) = p(\theta|x)p(x), \tag{C.13}$$

where $p(\theta, x) = p(x, \theta)$. Thus,

$$p(\theta|x)p(x) \quad = \quad p(x|\theta)p(\theta). \tag{C.14}$$

Dividing both sides of Equation C.14 by $p(x)$ yields Bayes' rule

$$p(\theta|x) \quad = \quad \frac{p(x|\theta)p(\theta)}{p(x)}. \tag{C.15}$$

Appendix D

Probability Density Functions

Histograms: If we measure the height of 5,000 people, and count the number of people with each height, then the resultant set of counts can be used to construct a *histogram*. A histogram is a graphical representation of a set of such counts, as shown in Figure 5.1a. Specifically, the set of $N = 5,000$ heights

$$\mathbf{x}_1 \quad = \quad (x_1, x_2, \ldots, x_N), \qquad (\text{D.1})$$

defines an ordered set of numbers, which can be used to construct a histogram as follows.

As we want to count the number of people with different heights, we need to divide the range of measured heights into a number of intervals, say, $\delta x = 1$ inch, between 60 and 84 inches (δ is the Greek letter *delta*). This yields a total of $N_x = 24$ intervals or *bins*, where each bin is defined by a lower and upper bound (eg the first bin spans 60-61 inches).

For each bin, we count the number of measured heights spanned by that bin. We expect relatively few measurements to be in the first and last bins, because they lie at the extreme ranges of human height. Conversely, we would expect a large proportion of measurements to be in bins at around 72 inches, because this is a common human height. The resultant histogram has a typical bell shape, shown in Figure 5.1a.

Of course the bin width does not have to be 1 inch, and we can define it to be the width δx, so the ith bin spans the range between x_i and $x_{i+1} = (x_i + \delta x)$. As before, the height of this bin is the number n_i of measured values that fall in this range. Thus, the heights of the bins

Probability Density Functions

in the histogram can be represented as

$$\mathbf{n} \;=\; (n_1, n_2, \ldots, n_{N_x}). \tag{D.2}$$

Note that Equation D.1 gives the value of N heights, whereas Equation D.2 gives the number of heights in each of N_x bins of a histogram.

Now, the total *area* A of all N_x bins represents the total number of measured heights. Thus, the area of a bin, expressed as a proportion of A, is equal to the probability that a randomly chosen value x falls within the range spanned by that bin

$$p(x \text{ is between } x_i \text{ and } x_{i+1}) \;=\; \frac{i\text{th bin area}}{\text{area of all bins, } A}. \tag{D.3}$$

The area of the ith bin is given by its height n_i times its width δx, (ie ith bin area $= n_i \times \delta x$), and the sum A of all N_x bin areas is therefore

$$A \;=\; \left(\sum_{i=1}^{N_x} n_i \right) \times \delta x, \tag{D.4}$$

where the capital Greek letter $\sum$ (sigma) represents the sum of all terms to its right (see Appendix B). Now Equation D.3 becomes

$$p(x \text{ is between } x_i \text{ and } x_{i+1}) \;=\; \frac{n_i \delta x}{\sum_j n_j \delta x} \tag{D.5}$$

$$\;=\; n_i/N. \tag{D.6}$$

Given that x must have some definite value, the total probability of all values of x must sum to one

$$\frac{1}{A} \sum_{i=1}^{N_x} n_i \, \delta x \;=\; 1. \tag{D.7}$$

If we define $p(x_i) = n_i/A$ then, substituting this into Equation D.7

$$\sum_{i=1}^{N_x} p(x_i) \, \delta x \;=\; 1, \tag{D.8}$$

and as this amounts to summing the proportions of all bin areas, it is self-evidently one. As this represents a histogram in which each column area is divided by the total area of the histogram, the histogram has an area of one. In this *normalised* histogram, the area of the ith column represents the probability $p(x$ is between x_i and $x_{i+1})$.

Probability Density: If we reduce the bin width δx so that it approaches zero then it is written as dx. And, if we consider a bin with sufficiently small width then the bin height can be assumed to be constant with a value of $p(x_i)$. Thus, the area of the ith bin is $p(x_i)\,dx$, which is the probability that x is between x_i and $(x_i + dx)$.

One important, but rather subtle, implication of this is that the value of $p(x)$ for a variable Θ is a *probability density*, and not a probability. We have already established that the probability of x being between any two values corresponds to the area of a bin. It follows that the height at any point corresponds to a probability density, because it must be multiplied by a bin width dx to obtain a probability, and the distribution of a continuous variable is called a *probability density function* or pdf, as shown in Figure 5.1c.

Histogram Areas and Integrals: As the bin width δx approaches zero, by convention, it is replaced by the symbol dx, and the summation symbol $\sum$ is replaced by the extended S symbol for *integration*

$$\int_x p(x)\,dx \quad = \quad 1. \qquad (\text{D.9})$$

The area under a probability density function is, by definition, one.

Averages From Distributions: Suppose we have a normalised histogram of X values, and we wish to find the mean value. If each bin width is $\delta x = x_i - x_{i+1}$ then the proportion of X values between x_i and x_{i+1} is $p(x_i)\delta x$. When considered over the N_x bins, the mean value of x is therefore

$$\text{E}[x] \quad = \quad \sum_{i=1}^{N_x} p(x_i)\,x_i\,\delta x, \qquad (\text{D.10})$$

where E is standard notation for the mean or *expected value*. As the bin width δx approaches zero, the summation over bins becomes an integral over values of x between $x = -\infty$ and $x = \infty$

$$E[x] \quad = \quad \int_{x=-\infty}^{\infty} p(x)\, x\, dx. \tag{D.11}$$

This can also be used to find the mean of a function $f(x)$ of x

$$E[f(x)] = \int_{x=-\infty}^{\infty} p(x)\, f(x)\, dx. \tag{D.12}$$

Note that this effectively provides a weighting $p(x)$ to the function value $f(x)$ associated with each value of x, where each weighting represents the relative prevalence of one value of x. Of course, this applies to any variable, such as θ, so that

$$E[f(\theta)] = \int_{\theta=-\infty}^{\infty} p(\theta)\, f(\theta)\, d\theta. \tag{D.14}$$

Evaluating The Quadratic Loss Function: The mean of the quadratic loss function (defined on page 100) can now be obtained by taking Equation D.14, and replacing $p(\theta)$ with the posterior pdf $p(\theta|x)$, and the function $f(\theta)$ with the quadratic loss function $\Delta^2 = (\theta - \hat{\theta})^2$

$$E[\Delta^2] \quad = \quad \int_{\theta=-\infty}^{\infty} p(\theta|x)\, (\theta - \hat{\theta})^2\, d\theta. \tag{D.15}$$

As above, this provides a weighting $p(\theta|x)$ to the loss $(\theta-\hat{\theta})^2$ associated with each value of θ, where each weighting represents the posterior probability of one value of θ.

Appendix E

The Binomial Distribution

Permutations and Combinations: When we flip a coin twice, we observe a particular sequence or *permutation*; for example, a head followed by a tail. In contrast, a head and a tail, in any order, is called a *combination*. A head and a tail can exist as one of two possible permutations $\mathbf{x}_{ht} = (x_h, x_t)$, or $\mathbf{x}_{th} = (x_t, x_h)$. The probability of observing each of these permutations is the same because the order in which heads and tails occurs does not affect their probabilities. Thus,

$$
\begin{aligned}
p(\mathbf{x}_{ht}|\theta) \quad &= \quad p(x_h|\theta) \times p(x_t|\theta) &\text{(E.1)} \\
&= \quad \theta \times (1 - \theta) &\text{(E.2)} \\
p(\mathbf{x}_{th}|\theta) \quad &= \quad p(x_t|\theta) \times p(x_h|\theta) &\text{(E.3)} \\
&= \quad (1 - \theta) \times \theta, &\text{(E.4)}
\end{aligned}
$$

and if we assume $\theta = 0.6$ then the probability of the permutations $p(\mathbf{x}_{ht}|\theta) = p(\mathbf{x}_{th}|\theta) = 0.24$. As each of these two permutations occurs with probability 0.24, it follows that the probability of observing a head and a tail in any order, or equivalently, the probability of a combination that contains a head and a tail, is

$$
p(\mathbf{x}_{ht}|\theta) + p(\mathbf{x}_{th}|\theta) \quad = \quad 0.48. \qquad\qquad \text{(E.5)}
$$

We can clean up the notation by defining $\mathbf{x}$ as a specific combination, such as $\mathbf{x} = \{x_t, x_h\}$, where this is read as *one head out of two coin flips in any order*, and where we use curly brackets to represent a combination. Thus, we can write

$$p(\mathbf{x}|\theta) \quad = \quad p(\{x_t, x_h\}|\theta) \qquad\qquad \text{(E.6)}$$

$$= \quad 2 \times \theta \times (1-\theta) \qquad\qquad \text{(E.7)}$$

$$= \quad 0.48, \qquad\qquad\qquad\qquad\quad \text{(E.8)}$$

where $\theta(1-\theta)$ is the probability that any permutation contains exactly one head and one tail, and 2 is the number of permutations that contain exactly one head and one tail.

We can reassure ourselves that Equation E.8 is true by working out the probability of observing the three possible combinations that can result from two coin flips, namely, $\{x_t, x_t\}$, $\{x_t, x_h\}$, and $\{x_h, x_h\}$, given a specific value for θ. The combination $\{x_t, x_t\}$ includes only one permutation (x_t, x_t), so

$$p(\{x_t, x_t\}|\theta) \quad = \quad p(x_t|\theta) \times p(x_t|\theta) = 0.16. \qquad \text{(E.9)}$$

Similarly, for the combination $\{x_h, x_h\}$, we have

$$p(\{x_h, x_h\}|\theta) \quad = \quad p(x_h|\theta) \times p(x_h|\theta) = 0.36. \qquad \text{(E.10)}$$

So, given that the only possible three combinations from two coin flips contain zero heads $\{x_t, x_t\}$, one head $\{x_h, x_t\}$, or two heads $\{x_h, x_h\}$, with probabilities

$$p(\{x_t, x_t\}|\theta) \quad = \quad 0.16 \qquad\qquad \text{(E.11)}$$

$$p(\{x_h, x_h\}|\theta) \quad = \quad 0.36 \qquad\qquad \text{(E.12)}$$

$$p(\{x_h, x_t\}|\theta) \quad = \quad 0.48, \qquad\qquad \text{(E.13)}$$

the sum of these probabilities is one

$$0.16 + 0.36 + 0.48 = 1. \qquad\qquad \text{(E.14)}$$

The general lesson to be drawn from this simple example is as follows. To find the probability of observing a combination that contains n heads and $N-n$ tails, first find the probability of one such permutation, and multiply this by the number of permutations that contain n heads and $N-n$ tails, where this number is a *binomial coefficient* (described next).

The Binomial Coefficient: The number of permutations that contain exactly x heads amongst N coin flips is

$$C_{N,x} = N!/(x!\,(N-x)!), \tag{E.15}$$

where $C_{N,x}$ is a binomial coefficient and is pronounced N *choose* x. If a coin is flipped $N = 10$ times and the number of heads is $x = 7$ then the number of permutations that contain $x = 7$ heads and $N - x = 3$ tails is $C_{N,x} = 120$. If a coin has a bias $\theta = 0.7$ then the probability of obtaining any one of these permutations, such as, $\mathbf{x}_7 = (x_h, x_h, x_h, x_h, x_h, x_h, x_h, x_t, x_t, x_t)$, is

$$p(\mathbf{x}_7|\theta) = \theta^7(1-\theta)^3 = 2.223 \times 10^{-3}. \tag{E.16}$$

Given that the probability of obtaining any other permutation with $x = 7$ heads and $N - x$ tails is also 2.223×10^{-3}, and that there are $C_{N,x}$ such permutations, it follows that the probability of obtaining x heads and $N - x$ tails in any order (ie a combination) is

$$p(x|\theta) = C_{N,x}\,p(\mathbf{x}_7|\theta) \tag{E.17}$$
$$= 120 \times (2.223 \times 10^{-3}) \tag{E.18}$$
$$= 0.267. \tag{E.19}$$

The Binomial Distribution: If we flip this coin N times then the probability of observing x heads given a coin bias θ is

$$p(x|\theta, N) = C_{N,x}\,\theta^x(1-\theta)^{N-x}. \tag{E.20}$$

The *binomial distribution* for $N = 10$ and $\theta = 0.7$ is shown in Figure E.1. (see Section 4.1, p80).

The binomial distribution is a cornerstone in the analysis of binary events. As the number N (eg coin flips) increases, given some mild

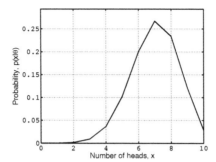

Figure E.1: The binomial distribution. Given a coin with a bias of $\theta = 0.7$, which is flipped $N = 10$ times, the probability $p(x|\theta)$ of obtaining different numbers x of heads defines a binomial distribution.

assumptions, the binomial distribution becomes increasingly like the Gaussian distribution (see Chapter 5 and Appendix F). Because the Gaussian distribution is mathematically convenient, this approximation allows the binomial distribution to be replaced with the Gaussian distribution in a wide variety of contexts.

Appendix F

The Gaussian Distribution

The equation for a Gaussian distribution is with mean μ_{pop} and standard deviation σ_{pop} is

$$p(x_i | \mu_{pop}, \sigma_{pop}) \quad = \quad k_{pop}\, e^{-(x_i - \mu_{pop})^2 / (2\sigma_{pop}^2)}, \tag{F.1}$$

where the constant $e = 2.718$, and $k_{pop} = 1(/\sigma_{pop}\sqrt{2\pi})$, which ensures that the area under the Gaussian distribution sums to one. For convenience, we define (minus) the exponent in Equation F.1 as

$$z_i \quad = \quad (x_i - \mu_{pop})^2 / (2\sigma_{pop}^2), \tag{F.2}$$

so that Equation F.1 can be written more succinctly as

$$p(x_i | \mu_{pop}, \sigma_{pop}) \quad = \quad k_{pop}\, e^{-z_i}, \tag{F.3}$$

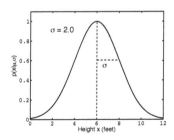

Figure F.1: A Gaussian distribution of heights x with mean $\mu = 6$ and standard deviation $\sigma = 2$, as indicated by the horizontal dashed line. This distribution has been re-scaled to a maximum value of one.

where $e^{-z_i} = 1/e^{z_i}$, which implies that, as z_i gets smaller, so $p(x_i|\mu_{pop}, \sigma_{pop})$ gets larger.

Let's begin by considering different values of x_i, assuming the mean and standard deviation are constant for now. If $x_i = \mu_{pop}$ then the square of the difference $(x_i - \mu_{pop})^2$ has a value of zero, at which point z_i adopts its smallest possible value (zero)

$$z_i = (x_i - \mu_{pop})^2/(2\sigma_{pop}^2) = 0. \qquad \text{(F.4)}$$

The constant $e = 2.718$ in Equation F.1 raised to the power $z_i = 0$ is one, so that, if $x_i = \mu_{pop}$ then $p(x_i|\mu_{pop}, \sigma_{pop})$ adopts its biggest value

$$p(x_i|\mu_{pop}, \sigma_{pop}) = k_{pop}\, e^{-z_i} = k_{pop}, \qquad \text{(F.5)}$$

where (as a reminder) $e^0 = 1$. As x_i deviates from μ_{pop}, so the exponent z_i increases, which decreases the value of $p(x_i|\mu_{pop}, \sigma_{pop})$. This decrease defines the fall-off in values on both sides of the peak. Thus, the peak of the Gaussian curve corresponds to the point at which $x_i = \mu_{pop}$.

Now, we consider the effect of altering the mean. Quite simply, because the mean determines the location of the centre of the distribution, changing its value moves the distribution along the x-axis. For this reason, the mean is known as a *location parameter*.

Next, we consider the effect of changing the standard deviation on the width of the distribution. If the standard deviation σ_{pop} changes then this effectively 'scales' the difference $(x_i - \mu_{pop})$. In order to see this, we first re-write Equation F.2 as

$$z_i = \frac{1}{2}\left(\frac{x_i - \mu_{pop}}{\sigma_{pop}}\right)^2. \qquad \text{(F.6)}$$

Now it is easier to see that increasing σ_{pop} has the same effect on the exponent z_i as decreasing the difference $(x_i - \mu_{pop})$. Thus, increasing σ_{pop} increases the spread of the bell curve. The standard deviation parameter σ_{pop} is known as a *scale parameter*, because it magnifies or shrinks the width of the Gaussian distribution.

Appendix G

Least-Squares Estimation

Here, we derive the least-squares estimate (LSE) of the true value μ_{true} of the mean μ. Given Equation 5.19 (repeated here),

$$E \quad = \quad \sum_{i=1}^{n} (x_i - \mu)^2, \qquad (G.1)$$

where we have replaced $-\log p(m, c|\mathbf{x})$ with E for simplicity. We will show that the minimum value of E is obtained when

$$\mu_{est} \quad = \quad (1/n) \sum_{i=1}^{n} x_i. \qquad (G.2)$$

At a minimum, the derivative of E with respect to μ is zero,

$$\frac{\partial E}{\partial \mu} \quad = \quad 0, \qquad (G.3)$$

where the *partial derivative* symbol ∂ is used instead of d when we take the derivative of only one of a number of possible parameters, whilst assuming that the other parameter values remain constant. Any book on elementary calculus tells us that all we need to do is to find an expression for this derivative in terms of μ, set it to zero, and then solve for μ.

The derivative of E with respect to μ is

$$\frac{\partial E}{\partial \mu} = \frac{\partial \sum_{i=1}^{n}(x_i - \mu)^2}{\partial \mu} \tag{G.4}$$

$$= \sum_{i=1}^{n} \frac{\partial(x_i - \mu)^2}{\partial \mu} \tag{G.5}$$

$$= -\sum_{i=1}^{n} 2(x_i - \mu). \tag{G.6}$$

Decomposing this sum into two components yields

$$\frac{\partial E}{\partial \mu} = \sum_{i=1}^{n} 2\mu - \sum_{i=1}^{n} 2x_i \tag{G.7}$$

$$= 2n\mu - \sum_{i=1}^{n} 2x_i. \tag{G.8}$$

The slope, and therefore this derivative, is zero at a minimum

$$2n\mu - \sum_{i=1}^{n} 2x_i = 0, \tag{G.9}$$

from which it follows that

$$\mu = (1/n) \sum_{i=1}^{n} x_i, \tag{G.10}$$

where μ is actually our estimated value μ_{est} of the mean based on n observations, so we can re-write this as

$$\mu_{est} = (1/n) \sum_{i=1}^{n} x_i, \tag{G.11}$$

which is the same as Equation 5.20. Thus, μ_{est} is the least-squares estimate (LSE) of the true value of μ.

Appendix H

Reference Priors

The question of what constitutes an un-biased or fair prior has several answers. Here, we provide a brief account of the answer given by Bernardo(1979)[3], who called them *reference priors*.

Reference priors rely on the idea of *mutual information*. In essence, the mutual information between two variables is a measure of how tightly coupled they are, and can be considered to be a general measure of the correlation between variables. More formally, it is the average amount of *Shannon information* conveyed about one variable by the other variable. For our purposes, we note that the mutual information $I(x, \theta)$ between x and θ is also the average difference between the posterior $p(\theta|x)$ and the prior $p(\theta)$, where this difference is measured as the *Kullback-Leibler divergence*. A reference prior is defined as that particular prior which makes the mutual information between x and θ as large as possible, and (equivalently) maximises the average Kullback-Leibler divergence between the posterior and the prior.

What has this to do with fair priors? A defining, and useful, feature of mutual information is that it is immune or *invariant* to the effects of transformations of variables. For example, if a measurement device adds a constant amount k to each reading, so that we measure x as $y = x + k$, then the mean θ becomes $\phi = \theta + k$, where θ and ϕ are *location parameters*. Despite the addition of k to measured values, the mutual information between ϕ and y remains the same as the mutual information between θ and x; that is, $I(y, \phi) = I(x, \theta)$. Thus, the fairness of a prior (defined in terms of transformation invariance) is guaranteed if we choose a common prior for θ and ϕ which ensures that $I(y, \phi) = I(x, \theta)$. Indeed, it is possible to harness this equality to derive priors which have precisely the desired invariance. It can be shown that the only prior that satisfies this equality for a location parameter (such as the mean) is the uniform prior.

As a more concrete example, suppose we wish to estimate the length θ of a table, based on many noisy measurements of x inches each. If we accidentally included a blank part at the beginning of the ruler, which has length k inches, then each measurement would be $y = x + k$ inches, and the mean would be $\phi = \theta + k$ inches. Whichever prior $p(\theta)$ we use for the mean θ of x, the corresponding prior $p(\phi)$ for the mean ϕ of $y = x + k$ should remain fair, regardless of the accidental offset k. As stated above, the only prior that guarantees this is the uniform prior.

As a further example, if a measurement device multiplies each reading by a constant c then we would measure x as $z = cx$. This could occur because the tape measure used is made of a material that has stretched (so that each measurement is in error by a constant proportion), or it could because we do not know whether lengths were measured in inches or feet. In either case, we would be ignorant of the scale of the measurements. If we define σ to be the standard deviation of x then the parameter σ would get transformed to $\psi = c\sigma$, where σ and ψ are *scale parameters*. As in the previous example, if a prior is fair for σ then it should remain fair for $\psi = c\sigma$, so that $p(\theta) = p(c\theta)$ for any value of c. It can be shown that the only prior that satisfies the equality $I(z, \psi) = I(x, \sigma)$ for a scale parameter is $p(\sigma) = 1/\sigma$, which is therefore the reference prior.

Of course, if it is obvious what the correct prior is then, for the sake of consistency, this prior should also be a reference prior. For example, for the joint pdf $p(x, \theta)$, the marginal pdf $p(\theta)$ is, by definition, the correct prior (see Chapter 6). Crucially, it can be shown that this marginal pdf of $p(x, \theta)$ is indeed the pdf that makes the mutual information between x and θ as large as possible, and is therefore the reference prior for θ.

References

[1] Bayes, T. (1763). An essay towards solving a problem in the doctrine of chances. *Phil Trans Roy Soc London*, 53:370–418.

[2] Beaumont, M. (2004). The Bayesian revolution in genetics. *Nature Reviews Genetics*, pages 251–261.

[3] Bernardo, J. (1979). Reference posterior distributions for Bayesian inference. *J. Royal Statistical Society B*, 41:113–147.

[4] Bernardo, J. and Smith, A. (2000). *Bayesian Theory*. John Wiley and Sons Ltd.

[5] Bishop, C. (2006). *Pattern Recognition and Machine Learning*. Springer.

[6] Cowan, G. (1998). *Statistical Data Analysis*. OUP.

[7] Cox, R. (1946). Probability, frequency, and reasonable expectation. *American Journal of Physics*, 14:113.

[8] Dienes, Z. (2008). *Understanding Psychology as a Science: An Introduction to Scientific and Statistical Inference*. Palgrave Macmillan.

[9] Donnelly, P. (2005). Appealing statistics. *Significance*, 2(1):46–48.

[10] Doya, K., Ishii, S., Pouget, A., and Rao, R. (2007). *The Bayesian Brain*. MIT, MA.

[11] Efron, B. (1979). Bootstrap methods: Another look at the jackknife. *Ann. Statist.*, 7(1):1–26.

References

[12] Frank, M. C. and Goodman, N. D. (2012). Predicting pragmatic reasoning in language games. *Science*, 336(6084):998.

[13] Geisler, W. and Diehl, R. (2002). Bayesian natural selection and the evolution of perceptual systems. *Philosophical Transactions of the Royal Society London (B) Biology*, 357:419–448.

[14] Gelman, A., Carlin, J., Stern, H., and Rubin, D. (2003). *Bayesian Data Analysis, Second Edition*. Chapman and Hall, 2nd edition.

[15] Geman, S. and Geman, D. (1993). Stochastic relaxation, Gibbs distributions and the Bayesian restoration of images. *Journal of Applied Statistics*, 20:25–62.

[16] Good, I. (1979). Studies in the history of probability and statistics. XXXVII A. M. Turing's statistical work in World War II. *Biometrika*, 66(2):393–396.

[17] Hobson, M., Jaffe, A., Liddle, A., and Mukherjee, P. (2009). *Bayesian Methods in Cosmology*. Cambridge University Press.

[18] Jaynes, E. and Bretthorst, G. (2003). *Probability Theory: The Logic of Science*. Cambridge University Press, Cambridge.

[19] Jeffreys, H. (1939). *Theory of Probability*. Oxford University Press.

[20] Jones, M. and Love, B. (2011). Bayesian fundamentalism or enlightenment? On the explanatory status and theoretical contributions of Bayesian models of cognition. *Behavioral and Brain Sciences*, 34:192–193.

[21] Kadane, J. (2009). Bayesian thought in early modern detective stories: Monsieur Lecoq, C. Auguste Dupin and Sherlock Holmes. *Statistical Science*, 24(2):238–243.

[22] Kersten, D., Mamassian, P., and Yuille, A. (2004). Object perception as Bayesian inference. *Ann Rev Psychology*, 55(1):271–304.

[23] Knill, D. and Richards, R. (1996). *Perception as Bayesian inference*. Cambridge University Press, New York, NY, USA.

[24] Kolmogorov, A. (1933). *Foundations of the Theory of Probability.* Chelsea Publishing Company, (English translation, 1956).

[25] Land, M. and Nilsson, D. (2002). *Animal eyes.* OUP.

[26] Lawson, A. (2008). *Bayesian Disease Mapping: Hierarchical Modeling in Spatial Epidemiology.* Chapman and Hall.

[27] Lee, P. (2004). *Bayesian Statistics: An Introduction.* Wiley.

[28] MacKay, D. (2003). *Information theory, inference, and learning algorithms.* Cambridge University Press.

[29] McGrayne, S. (2011). *The Theory That Would Not Die.* YUP.

[30] Migon, H. and Gamerman, D. (1999). *Statistical Inference: An Integrated Approach.* Arnold.

[31] Oaksford, M. and Chater, N. (2007). *Bayesian Rationality: The probabilistic approach to human reasoning.* Oxford University Press.

[32] Parent, E. and Rivot, E. (2012). *Bayesian Modeling of Ecological Data.* Chapman and Hall.

[33] Penny, W. D., Trujillo-Barreto, N. J., and Friston, K. J. (2005). Bayesian fMRI time series analysis with spatial priors. *NeuroImage*, 24(2):350–362.

[34] Pierce, J. (1961 reprinted by Dover 1980). *An introduction to information theory: symbols, signals and noise.* Dover (2nd Edition).

[35] Reza, F. (1961). *Information Theory.* New York, McGraw-Hill.

[36] Rice, D. and Spiegelhalter, K. (2009). Bayesian statistics. *Scholarpedia*, 4(3):5230.

[37] Simpson, E. (2010). Edward Simpson: Bayes at Bletchley park. *Significance*, 7(2).

[38] Sivia, D. and Skilling, J. (2006). *Data Analysis: A Bayesian Tutorial.* OUP.

[39] Stigler, S. (1983). Who discovered Bayes's theorem? *The American Statistician*, 37(4):290–296.

[40] Stone, J. (2011). Footprints sticking out of the sand (Part II): Children's Bayesian priors for lighting direction and convexity. *Perception*, 40(2):175–190.

[41] Stone, J. (2012). *Vision and Brain: How we perceive the world*. MIT Press.

[42] Stone, J., Kerrigan, I., and Porrill, J. (2009). Where is the light? Bayesian perceptual priors for lighting direction. *Proceedings Royal Society London (B)*, 276:1797–1804.

[43] Taroni, F., Aitken, C., Garbolino, P., and Biedermann, A. (2006). *Bayesian Networks and Probabilistic Inference in Forensic Science*. Wiley.

[44] Tenenbaum, J. B., Kemp, C., Griffiths, T. L., and Goodman, N. D. (2011). How to grow a mind: Statistics, structure, and abstraction. *Science*, 331(6022):1279–1285.

Index

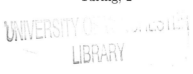

About the Author. Dr James Stone is a Reader in Computational Neuroscience at the University of Sheffield, England.

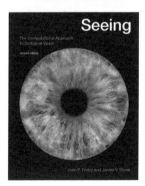